OIL UNDER THE ICE

Cover design: BB&H Graphic Communications Ltd.

ISBN 0-919996-01-9

Printed and bound in Canada, 1976
Dollco Printing Limited
525 Coventry Road
Ottawa, Ontario
K1G 3M5

Canadian Arctic Resources Committee
1976 Publishing Programme
Edie Van Alstine, Series Editor

Oil Under the Ice

Douglas H. Pimlott
Dougald Brown
Kenneth P. Sam

CANADIAN ARCTIC RESOURCES COMMITTEE
46 ELGIN STREET, ROOM 11
OTTAWA, ONTARIO
K1P 5K6

To Sam Raddi

Canadians have now become intensely aware of the North. The concept of the last frontier is no longer a play on words; we now recognize that the North is a region of the country that we have the opportunity to develop in special ways; we recognize that if it is developed carefully and wisely it could play a powerful role in the development of our culture; we recognize that it could greatly alter our dependency on the culture, the markets, and the technology of other countries. We feel very strongly that its potential for moulding our nation, its potential to provide young Canadians with a region of their own, must not be lost by precipitous development which could result in both social and environmental disaster.

Memorandum on the formation of the
Canadian Arctic Resources Committee, 30 April 1971.

CONTENTS

Foreword

This book is about risk-taking. The possibility of discovering huge reservoirs of oil and gas in the Arctic set the largest corporations in the world, the United States' and Canadian governments into a frenzy of risk-taking. They had their own stakes and could win or lose fortunes. Others in society, like native people and environmentalists, were not allowed a place around the table. Yet if these expensive Arctic ventures failed they would bear the consequences. That is what offshore drilling in the Canadian Arctic is all about.

Without access to the discussion table, and without knowing where it would lead, the Canadian Arctic Resources Committee took risks of its own. We ventured into some blind alleys but also down paths which warranted travelling. *Oil Under The Ice* is the result of one of these forays. The Committee's involvement in offshore drilling dates back to our first conference in 1972, when CARC was asked to provide resource workers for northern organizations. The group under the most pressure at that time was the Committee for Original Peoples' Entitlement, a small and dynamic organization representing native people in the Mackenzie Delta-Beaufort Sea area. We did not know what kind of work was involved or what experience would be needed. Douglas Pimlott and I looked for six months and failed to identify the right person. We were about to give up when I realized that the man who could do the job had been in front of me all the time. I urged Pimlott to accept the challenge.

The Resource Worker Programme involved a wide variety of tasks. First, you had to be accepted by the native people. This acceptance takes time but Pimlott had the ingredient shared by many of his northern friends: a sense of time. While I sat in Ottawa waiting for immediate results I started receiving letters with comments like the following: "As you can imagine it just isn't possible to walk into a settlement and say, 'here I am you lucky people — what do you want me to do?' The problem is that the concept of a resource person is a rather abstract one — even to me. I thought that ideas and approaches would come out in conversation. Even that takes time to develop." (Tuktoyaktuk, November 12, 1973.) A week later I received this letter from Holman Island. "Life is very hard here. I have been forced for the past week to eat caribou, ptarmigan and Arctic char at least two meals a day. Oh how my mouth waters for a hotdog!!!" And further, "Jobs for a resource person promise to be quite varied. So far I have written a will, a set of papers for an old-age pension supplement and offered advice to a research officer on aboriginal rights for an Australian Royal Commission." Not very much to tell the CARC Board.

Pimlott soon learned of the imminence of wide scale offshore drilling operations. The plans were closely guarded by both industry and government, but he discovered that drilling in the deep waters of the Beaufort Sea was slated to begin in 1975 and that drilling from ice islands in the Sverdrup Basin would start in early 1974. No one realized the full implications of what was planned. Sam Raddi, president of COPE, asked Pimlott to investigate the situation and prepare a report. The research led to the report "Offshore Drilling in the Beaufort Sea", which he presented to COPE in January 1974. But by then there were other jobs to do and nothing more could be done on offshore drilling. When he returned south after a year in Inuvik he continued the original investigation that has finally led to *Oil Under The Ice*.

Almost every project that CARC has undertaken began on a shoe string. The Resource Worker Programme was no exception. We sought support from both the petroleum industry and the federal Government, because both reiterated their concern for native people. But one after another, 12 member companies of the Arctic Petroleum Operators' Association gave us their version of the familiar theme: "where two elephants are involved there is no room for a mouse". We first heard these words from the chairman of Imperial Oil in 1971, just after CARC was formed. The Committee was always told by the two elephants that we could play no role in northern affairs. Fortunately, the private foundations in Canada did not agree. In some of them there are people well aware of the dangers of wholesale reliance on the state. With their meagre resources they have attempted to redress the balance; CARC could not have done its work without their help. Although I have become pessimistic when I see how little Ottawa cares for the working of free self-government, it's heartening that some foundations have taken up the challenge.

The attempt to stifle initiative and harp on the inadequacy of individuals and small groups is widespread in Canada. In every area of private endeavour the federal Government seems intent upon crushing independent inquiry. This is taking its toll across the country. In my opinion, the voluntary and public-interest sector is either close to extinction or moving towards complete reliance on the state. This situation has in part been brought about by an alliance representing an unprecedented concentration of economic and political power between the federal Government and the petroleum industry. The two elephants joined forces several years ago and have since spent millions of dollars telling Canadians of the immediate need for a Mackenzie Valley gas pipeline. They tied up most of the country's northern experts on lucrative contracts. Many are still on contracts and remain effectively silenced. Don't misunderstand me; no one was forced into silence. On the contrary, the experts silenced themselves in one of the biggest shows of submission to corporate and government strength we have seen in years. The vitality of Canada's universities and scientific centres has suffered immeasurably.

The attempts to stifle plurality in this country must be fought. During the past few years CARC has tried to bring an objective voice into the dialogue now reserved for the energy giants and the federal and provincial governments. *Oil Under The Ice* is one of our efforts. We publish the book with full knowledge of its limitations. In spite of these limitations we think it shows that, even in the case of vast and complicated developments located in remote regions of our country, the citizens and their organizations, if they have the will, can be heard and can have a meaningful influence on public affairs.

Douglas Pimlott was the first Chairman of the Canadian Arctic Resources Committee. At the end of his first major address in 1972 he wrote, "But to return to the North, I feel certain that some things will not change. One of these is the increasing interest of people in the south for the place of the people who were there first. Another is an increasing awareness of the need to protect its fragile environments. One of the unanswered questions is whether people will be encouraged to shape events or not. Is there a reasonable alternative? Time will tell. Eventually people and the North will know."

For the past three and a half years he has singularly devoted himself to the task of shaping events. This book is one more tribute to that remarkable man's persistence.

Kitson Vincent
Executive Secretary
Canadian Arctic Resources Committee
January 1976.

Preface

It has been said that where men are involved there is no absolute truth. This account of the frantic exploration for oil and gas in Canada's Arctic has been mainly written by an ecologist/naturalist who has also written on the rights of wolves, on the Canadian wilderness, and on the real harm of the seal hunt. These papers reflect some of my attitudes, my perspectives and biases, and indicate that my interpretation of the events and facts on which this book is based is quite different from one which might come from the president of Panarctic Oils had he written about the same topic.

I am not making an apology for the perspective of this book. I think that some of my biases, for example that wolves have as much right to a place in the world as people, are survival "good sense" for the human race. We live in an ecosphere just as wolves do, and ecosystems that keep their wolves are usually in a lot healthier state than those that have lost them. I am no "expert" on the North, but because I worked there and experienced a little of the Arctic I couldn't keep from being involved in its problems.

I saw plants underlain by permafrost for the first time in June 1966. It was at an abandoned Dew Line site on Nadluardjuk Lake. My students and I used Fox Bravo as a base camp for our studies on wolves and caribou for the next five years. My awareness of man's impact on the Arctic environment developed gradually as I talked, read, and thought about the things I saw and experienced in that beautiful but austere land. Fox Bravo, which was abandoned in 1964, is on one of the highest hills in the vicinity. From it we could see radar domes of Fox 2, the station 55 miles to the west, and when we lowered our vision we could see a small mountain of oil drums to the north. To the east there was a radar tower which had broken into fragments when it crashed down the hillside after the guy wires had been cut. At one end of the airstrip there was a mound of snowmobiles, trucks, and tractors that had been put to the torch, apparently the last act before the site was abandoned. The largest bulldozer had been used to batter the other vehicles into immovable heaps of junk, then the bulldozer had been driven to the top of the mound, the caps were removed from drums of diesel fuel strapped to it, and the pile set alight.

The shore of Nadluardjuk Lake is dotted with petroleum drums, some high and dry, some partially submerged. When we examined them we found that a considerable number contained either diesel fuel or gasoline; they will pose a threat to the wild things of the area for many years because they will sooner or later rust out and release their deadly contents into the lake.

The bits and pieces of the Arctic jigsaw puzzle began to fall into place for me in 1969, when I served as a member of a study group on fisheries and wildlife for the Science Council of Canada. A study of the Arctic literature, the Tundra Conference in the fall of 1969, and contacts with other scientists demonstrated how little was known about the ecology of northern ecosystems and about methods of protecting the environment from harsh use or from pollution. At the same time, I became aware of the tremendous economic forces which had moved into the Arctic to find and develop its resources. I also learned that the legal framework and the administrative system for the protection of Arctic environments was virtually non-existent — eight years *after* the promulgation of the Oil and Gas Land Regulations, which had spurred exploration for oil in the North, and two years after the formation of Panarctic Oils had been stimulated by the infusion of capital from the federal Government.

As more pieces of the puzzle fitted together I realized that Fox Bravo was a precursor for the Canadian North in the 1970s, unless drastic changes took place in government and industry attitudes. It was then, early in 1971, that I decided to become actively involved and join other Canadians who were already at work trying to influence the inevitable processes which had left battered and burned vehicles mouldering on the tundra; which abandoned drums of diesel oil in Nadluardjuk Lake; which spurred exploration for oil and gas years before processes of environmental protection had been established, and years before industry had the technological capability to explore with the degree of safety that the environment and the interests of the native people warrant.

My understanding of inevitable processes in the development of non-renewable resources has greatly developed during the past two years as I have attempted to piece together the story of offshore drilling in the Arctic. Our objective in this book has been to show specifically how the federal Government has worked to implement its northern development policies. These policies state that the interests of the native people and the protection of the environment have a higher priority than the development of non-renewable resources. The facts however, are compelling evidence that this is not so — and that the policy statements have been simply empty rhetoric.

I am grateful to my wife Dorothy who accepted my absence from our home for almost a year when I worked with COPE in Inuvik. It was not my first absence for an extended period but it was one of the longest. That period with native people in the Mackenzie Delta and Beaufort Sea was one of the most rewarding, stimulating and educational periods of my life, although it was a lonely period for her.

Douglas H. Pimlott

Acknowledgements

Many people assisted during the two years involved in researching and writing this book. Members of the Committee for Original Peoples' Entitlement (COPE) were involved at the outset. Sam Raddi, COPE's president, gave initial direction to undertake the project. He, Nellie Cournoyer, Jean Look, and Jennifer Rigby were the ones who assisted, listened to, and encouraged during the frustrating days spent searching for information to write the first COPE report on offshore drilling in the Beaufort Sea.

Kit Vincent, Executive Secretary of CARC, was the first to suggest that the study should be extended beyond the initial Beaufort Sea project to all areas of Arctic waters. He also assisted by hounding us to meet deadlines. Marion Hummel attended to the infinite details in the latter stages. Edie Van Alstine served as series editor. H.T. Murphy edited the final manuscript. In the shadows there are a considerable number of individuals in government and industry who cannot be named. They provided information, insights, clues to documents and reports and in some cases, access to them. We express our appreciation and gratitude to all of those mentioned and to numerous others who helped in one way or another but who are not mentioned.

We also express our appreciation to Tom Smith, George Greene, and Don Mackay who wrote the papers on the biology of the Beaufort Sea and on oil spills and contingency plans respectively for inclusion as an appendix. Their papers add depth to two important areas which are dealt with only lightly in the text.

The work involved in the production and publication of this book was supported directly or indirectly by a number of organizations. The initial investigation was one of the Resource Worker projects conducted for COPE and was supported by the Richard and Jean Ivey Fund and the Laidlaw Foundation. The work was completed as a case history project of the Northern Resource Policy and Land Use Study and was supported by the Bronfman Foundation, the Molson Foundation, and by another Montreal foundation. COPE supported the work on the transportation of oil and gas from offshore. The Northern Environment Foundation provided funds towards publication.

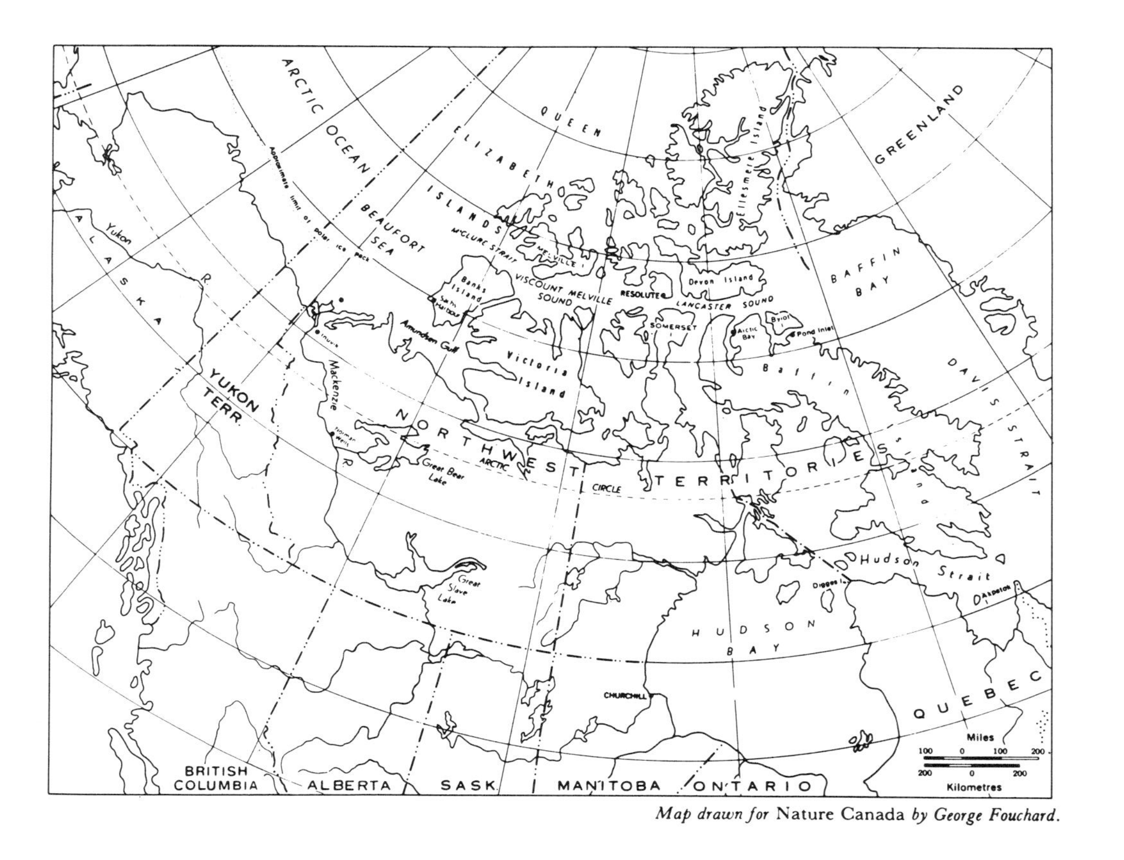

Map drawn for Nature Canada *by George Fouchard.*

Chapter 1

Oil Under the Ice: A Perspective on Offshore Drilling

Plans to drill for gas and oil in the offshore regions of the Canadian Arctic developed quietly for more than a decade. The offshore operations both proposed and underway represent a wide spectrum of drilling technology: artificial islands, monopods, and drilling barges in the Beaufort Sea; conventional drilling rigs on ice in the Arctic Archipelago; semi-submersible rigs in Hudson Bay and dynamically positioned drillships in Lancaster Sound.

As planning progressed both government and industry moved to prevent the Inuit of the N.W.T. and the Canadian public from becoming aware that the search for oil and gas was entering an entirely new and dangerous phase. On the government side, the coverup included granting permits for construction of artificial islands under land-related rather than water-related legislation so they appeared as simple extensions of the land rather than a completely new exploration frontier. Later, when programmes were formulated for operations which could not be so disguised, a thick blanket of secrecy was thrown over operations planning in both the Beaufort Sea and the Arctic Islands.[1]

Industry too did its best to keep information on plans for offshore drilling from becoming public knowledge. The high cost of developing drilling systems for offshore areas brought those companies with offshore interests together to form the Arctic Petroleum Operators' Association (APOA), thus allowing the industry to present a united front to public interest and native groups. But not until the Committee for Original Peoples' Entitlement (COPE) and CARC had given wide publicity to drilling plans did the annual APOA environmental conference deal with offshore drilling. In the Arctic Islands, Panarctic, the federal government-dominated consortium, kept the fact that it was drilling its first offshore well from ice in the High Arctic a close secret until a select group of reporters were invited to visit the site.[2]

The role of public-interest organizations in informing the public about social and environmental aspects of resource development programmes, and about the programmes themselves, is one of the themes of this book. The COPE/CARC reports[3] changed the situation from one in which not even reporters or members of Parliament knew about plans for offshore drilling, to one in which members of the general public became aware of the hazardous search for oil and gas in Arctic waters. When further secrecy was futile, the industry developed an information programme for native people in the vicinity of the Beaufort Sea. The limitations of the programme, however, made it necessary for native and public-interest groups to continue their own enquiries.[4]

Industry and government secrecy was paralleled by failure to establish research programmes to provide the basis for assessment of the environmental impact of offshore operations, the understanding to minimize the effect of operations, and the know-how to deal with the oil spills which will inevitably occur. Finally, when a research programme was established for the Beaufort Sea it was established as a crash programme which, at best, could provide only limited understanding of that complex and variable ecosystem. In other areas of the Arctic even less was done: no substantial research was undertaken, and the environmental impact assessment process turned science into the petroleum industry's tool.[5]

The adequacy of jurisdictional and administrative arrangements for protecting Arctic marine environments is questioned in almost every chapter of this book. A key issue, which the Canadian Arctic Resources Committee has raised many times since its formation in 1971, is the way Environment Canada (DOE) was relegated to a minor advisory role in environmental protection in the N.W.T. and the Yukon.[6] In retrospect it is evident that the policy enunciated in the 1970 Speech from the Throne was misleading and misrepresented the Government's intentions. The general policy was that the intention in forming the department was to resolve the "Inherent conflict of interest . . . between those who are seeking the exploitation of non-renewable resources and those who are charged with the responsibility of protecting the environment. This conflict is not irreconcilable, nor is there anything evil in it. Nevertheless, the Government is of the opinion that until such time as environmental values are firmly entrenched those differences are better debated and resolved by ministers in council and not by officials in any department."[7]

Also in 1970, the Government passed the Northern Inland Waters Act and the Arctic Waters Pollution Prevention Act, and did not give Environment Canada even a joint mandate to administer them. DOE's role was progressively reduced so that it could no longer enforce Section 33 (the anti-pollution section) of the Fisheries Act without consulting the Department of Indian and Northern Affairs (DINA), the department which has always marched to industry's drum.[8]

Since 1969, when DINA made the first major move to gain complete control of northern environmental affairs, a flood of promises came from the department about protection of the environment. They were made in numerous ministerial speeches and in other documents including a Government policy statement prepared for Cabinet in 1971 and approved in 1972.[9] But the insincerity of the speeches and the policy statement is evident when measured against the approach taken to promote the development of offshore drilling, and against the enforcement of environmental regulations. It is obvious that there is no environmental philosophy, and in its absence no meaningful environmental policies will be implemented by DINA.

But one cannot be certain that the situation would have been very different if DOE had administered the environmental legislation for the Arctic. The Environmental Assessment Review Process (EARP) which DOE established was one means by which it might have forced compliance with strict environmental standards, but it was not given any legislative base and appeared to be an administrative mechanism that could be easily bypassed. There did not seem to be much possibility that it would reduce the environmental hazards of offshore drilling.[10] It is difficult to assess with certainty the relative importance of the various players: how important was it that DOE has had weak, or disinterested, Ministers who would not fight hard battles in Cabinet? how important was it that the Prime Minister and his Cabinet are biased toward exploitation of non-renewable resources? how important was it that two groups of senior administrators, the Advisory Committee on Northern Development and the Task Force on Northern Oil Development, had strong pro-development sympathies and close contacts with the mining and petroleum industries? how important was it that DINA had an Assistant Deputy Minister for Northern Affairs who was a highly skilled bureaucratic politician? Logic rules against the idea of a simplistic answer. It is far more likely that all of these factors, and still others, were part of a complex political and bureaucratic pattern.

Cooperation between government and industry to maintain secrecy, disregard for the interests of the native people, the use of inadequate drilling systems and the triumph of interdepartmental politics at the expense of environmental protection, are all part of the pattern. The rush to drill in the Arctic Ocean is not simply a matter of the simplistic application of technology to obtain energy to fuel southern industry and heat southern homes. Rather, it is an issue of public policy that raises far deeper questions. Perhaps the most fundamental question is the competence of present government structures to manage the interface between the application of unproven technology and the protection of frontier environments in the Arctic. The rush to drill represents the leading edge of a wave of demands for non-renewable resources which could overwhelm Inuit culture in the North as Indian cultures have been overwhelmed in the South. Sadly, the evolution of public policy for the Arctic Ocean is a story of the ascendancy of bureaucracy over democratic process and of secrecy over public participation and involvement.

Setting the Stage

The hard-line development of oil and gas in Arctic offshore areas is not unique. The same approach characterized mining in the Yukon and N.W.T., promoting oil and gas pipelines through the Mackenzie Valley, and exploration for petroleum on land. During the past 20 years the federal Government has actively encouraged the multinational resource corporations to explore for and develop non-renewable resources north of the 60th parallel.[11] Offshore drilling is simply one facet of a northern policy which was articulated by John Diefenbaker almost 20 years ago. The election of 1958 focused on what a Conservative Government would do to promote resource development. Diefenbaker's "Vision of the North" was a prominent Conservative Party campaign slogan. Later events show clearly that the stage was being set for a programme of intensive development. Land and geological surveying was accelerated, and in 1960 improved geological maps were published. These maps showed the location of formations likely to bear petroleum and minerals. Prior to 1964, the average number of claims staked north of 60 was less than 6000 a year. During the next five years there were five major staking rushes, reaching a peak of 52,000 claims staked in 1968 and representing a high level of activity in exploration and development.[12]

In 1964, a railway was completed to Hay River on Great Slave Lake and to Pine Point where a large lead-zinc mine was being developed. Late in the decade intensive studies were conducted to determine the feasibility of establishing a deepwater port at Herschel Island on the Beaufort Sea. In 1966, a Northern Exploration Assistance Programme was established which provided up to 40% of the exploration costs of Canadian citizens, or of companies incorporated in Canada. Norman Wells, discovered in 1918, was the only oil field in production before the '60s. Although the acreage held by oil companies under exploratory permits increased rapidly after the publication of the geological maps in 1960, much of the interest was of a speculative nature and there was little exploration of permit areas in the early part of the decade.

The announcement of a major discovery of oil and gas at Prudhoe Bay came just after the Canadian Government had joined a consortium to form Panarctic Oils, in which it held 45% of the equity. The new company immediately began an active exploration programme on its holdings of approximately 44 million acres in the Arctic Islands, the most successful operation in northern Canada. It has discovered approximately 15 trillion cubic feet of gas and undisclosed quantities of oil. But at the same time Panarctic's record in the conduct of its operations is unimpressive. It made two major gas discoveries out of the first eight wells drilled. Both blew out, one on Melville Island and one on King Christian Island. Fortunately for the environment, neither discovery was crude oil.[13]

Many of the large multinational corporations also became interested in the potential of the Arctic, and by 1971, 460 million acres were under permit. Earlier, the oil industry had been invited to draft the oil leasing regulations so they would provide maximum incentives for exploration. The leasing arrangements were much more favourable than those which prevailed in Alaska and in terms of petroleum speculation they certainly accomplished their purposes.[14] The areas held under exploration permits represented virtually all the potential oilbearing formations north of the 60th parallel. They also included large offshore holdings, covering the Beaufort Sea to the edge of the polar pack and virtually all the waters enclosed by the Arctic Islands including the Northwest Passage, and even M'Clure Strait at the northwest end of the Passage, normally free of ice only one year in ten. In the post-Prudhoe Bay climate, expenditures on exploration increased from $24 million in 1967 to $170 million in 1971 and remained high through 1974 (Tables 1 and 2).[15]

There have been several discoveries of gas and oil in both the Mackenzie Delta region of the western Arctic as well as in the Arctic Islands. The combined total of all gas finds still falls short of that discovered in the Prudhoe Bay reservoir. Although no announcements have been made on the size of the oil discoveries, it is believed that their volume still falls far below that required for commercial production. Today much of the focus in exploration is changing from the land to the sea. It appears that the results of exploration programmes have indicated that major land discoveries are unlikely; the industry hopes that offshore regions contain reservoirs of the magnitude of Prudhoe Bay, and that is why they are eager to begin drilling wildcat wells in the Beaufort Sea in 1976.

During the 1960s few Canadian conservationists outside the federal Government realized the extent of the developments in the Arctic. Almost no questions were raised in Parliament by the press or even by the publications of conservation organizations about the potential impact of exploration on Arctic ecosystems. Canada had neither the individuals nor the organizations to take the initiative to alert the country to the environmental dangers. No one even challenged the Government to consider alternate forms of resource development in the decade after the "Vision of the North".

The Canadian Wildlife Service was the principal government agency interested in the protection of Arctic ecosystems. In the postwar period, members of CWS were actively engaged in studies of Arctic foxes, caribou, musk-oxen, waterfowl, seabirds, and other species, but most of these studies were over by the early 1960s. Local government was evolving in the Yukon and N.W.T. Game management was assumed by the Game Branches when territorial governments were formed. Communications between these two organizations were not good. CWS encountered strong resistance to its request for funds for Arctic research, and during the critical pre-development period, CWS had little influence on the formulation of environmental policies for the Arctic.

Responsibility for the aquatic ecosystems of the North came directly under the Department of Fisheries (which evolved into the Department of the Environment in 1972). There were no jurisdictional conflicts involved with either territorial government but the department was ineffective in representing Arctic environmental interests. The Fisheries Research Board (FRB) established an Arctic Biological Station in 1955, to be responsible for "Canadian research on marine and anadromous fish and biological oceanography in northern waters, as well as marine mammals on all coasts."[16] In terms of the two territories alone, its area of responsibility represented 30% of the area of Canada, but until 1972, the full-time scientific staff of the station had never exceeded nine. The primary interest of the department and of FRB was the commercial fisheries of the two coasts; as a result it did even less than the CWS to develop environmental awareness among the agencies who were so actively promoting the development of oil, gas, and minerals in the Arctic.

Alternatives, Options, and Offshore Drilling

Several chapters in this book conclude with questions about the wisdom of drilling offshore wells in situations where oil might blow out into the Arctic Basin for as long as a year before it could be stopped. Particularly in the case of Lancaster Sound, consideration of a no-drill option is urged. Risks of the magnitude involved here should have caused the Government to weigh the alternatives and options open to it before approving the programme, but the Cabinet memorandum on offshore drilling (Appendix 1) indicates that research on the possible hazards was very superficial. Even though large sums of money have already been invested in exploration and development, it is still valid to ask, should Canada be attempting to develop its Arctic petroleum resources now? more particularly, in the absence of drilling technology which is capable of coping with Arctic ice for more than three months of the year, should wildcat wells be drilled in deep-water areas of the Beaufort Sea, Sverdrup Basin, Northwest Passage, Hudson Bay, and Davis Strait?

There are many alternatives to conventional methods of petroleum exploration to supply our oil needs, for example, making a large-scale investment in development of the Athabasca tar sands instead of constructing two pipelines to the Arctic. Even if it were decided to develop Arctic resources, there would still be the alternative of concentrating on the development of either the eastern or western Arctic now, and the other area later. (There seem to be frightening economic implications in the multiplication of extensive short-term resource projects; this aspect of the question warrants national consideration also.)

If Arctic offshore resources must be developed in the immediate future, then government and industry should together develop advanced technology capable of operating in Arctic ice throughout the year. Industry at present appears to believe itself capital deficient, and so is very reluctant to invest in the research and development of technological innovations which would really allow it to cope with Arctic ice.[17]

Even if offshore petroleum resources were to be developed, there would remain the option of no development in some critical areas, such as Lancaster Sound.[18] Ecologically, the Sound is possibly the richest, most productive area in the Arctic. It has been proposed by the International Biological Program (IBP) as the site of a large ecological preserve. Because of the Sound's long-term value in producing animal resources for Inuit society and because of its potential scientific value, the decision-making process by which DINA gave its approval in principle for drilling there was completely inadequate, illustrating that a more socially conscious approach to evaluating energy needs and options is necessary.

Besides considering alternative ways of developing petroleum resources for energy purposes there are the long-term aspects of environmental degradation which must be, but have not been, examined: "Energy consumption is clearly at the base of almost every important environmental problem, from strip mining to unsightly hydro reservoirs, to transmission lines, to radioactive wastes, to oil spills and aesthetic desecration of the Arctic. Furthermore, pollution is an inevitable result of energy consumption, not a byproduct as we are so often led to believe."[19]

Even the Shah of Iran has been quoted as saying that oil and gas are too valuable to burn. His implication was, of course, that in the long term they are much too important to civilization to use recklessly to fuel our present extravagant life-style. The option of using Arctic petroleum resources to meet long-term petrochemical needs warrants urgent consideration; with its controlling interest in Panarctic Oils, the federal Government is in a strong position to buy out other member companies and to hold Arctic Island resources in reserve, should that option be favoured.

Looking into the future, it is difficult not to return to the ecologist-naturalist viewpoint and the words of Dr. Douglas Clarke: "The last area of values related to wildlife is the national heritage — certainly the most important. The national image of the North is of a vast land, but not an empty one. There are bears and wolves, caribou and muskoxen, geese and ptarmigan. The fact that Canada should ever be known as a land that failed to take due account of its wildlife in its haste for development is totally unacceptable to our people. We must be held responsible. How much is it worth to be proud? How much does it cost to be ashamed? These are the intangibles that we must evaluate when we make decisions."[20] It would be easy to brush off the idea that wildlife values should be given much weight in the development of energy options for the future but these values are important to human civilization, as well as for the animals which have long shared the Arctic with men and provided sustenance for them.

The native Inuit culture of the Arctic is deeply rooted in the ways of the animals, the land, and the sea. Even though the traditional life has been drastically changed, the Inuit are still very much dependent on the land and its resources for food and for spiritual well-being. This has never been a significant consideration when we, white society, established priorities for northern economic development projects. But their interests should weigh heavily in the balance when options for the development of oil and gas are being pondered.

Table 1

Northern Exploration Costs[15]

(NWT, Yukon, and Arctic)

(Millions of Dollars)

Year	Geological and Geophysical	Exploratory Drilling	Development Drilling	Other	Total
1975*	100.0	170.0	5.3	5.2	280.5
1974*	120.0	205.0	6.3	6.2	337.5
1973	92.3	158.7	5.6	5.5	262.1
1972	84.2	135.0	1.8	5.5	226.5
1971	80.4	80.5	1.1	4.4	166.4
Cum. 1947-1970	195.5	153.6	9.2	56.7	415.0
TOTAL	672.4	902.8	29.3	83.5	1688.0

*Oilweek estimate.

Table 2

Northern Drilling Activity[15]

(1968 — 1975)

Year	N.W.T. and Yukon		Arctic Islands	
	Completions	Footage	Completions	Footage
1975*	28	208,000	12	120,000
1974	37	295,698	23	210,265
1973	60	447,279	23	198,941
1972	51	395,970	20	170,333
1971	62	355,779	14	113,508
1970	67	319,471	6	45,260
1969	53	252,219	2	10,070
1968	39	131,877		
TOTAL	387	2,406,293	100	868,377

Source: Department of Indian and Northern Affairs
**Oilweek* estimate.

References and Notes

1 See chapters 2 and 3 on the Beaufort Sea, and chapter 5 on the Arctic Islands.

2 See the reference to government and industry as two elephants and to public interest organizations as a mouse, made by a petroleum industry official to a CARC fund-raising group in 1972. It became the theme of an article, *Two elephants and a mouse,* K. Vincent, which appeared in *Nature Canada,* July/September 1972.

3 Committee for Original Peoples' Entitlement, *Drilling for oil and gas in the Beaufort Sea,* press release, 8 February 1974, and the background report *Offshore drilling in the Beaufort Sea,* D.H. Pimlott; also two issues of *Northern Perspectives,* vol. 2 no. 2 and vol. 2 no. 4, 1974, D.H. Pimlott, *The hazardous search for oil and gas in Arctic waters, Nature Canada,* October/December 1974.

4 See chapter 2 for a discussion of the way public information processes worked for offshore drilling in the Beaufort Sea.

5 See chapter 5 for a discussion of the way the environmental impact assessment process worked for offshore drilling in the Arctic Islands.

6 Pimlott, D.H., *People and the North; motivations, objectives and approach of the Canadian Arctic Resources Committee, Arctic Alternatives* (Ottawa: CARC, 1973).

7 Canada, House of Commons, Debates, 3rd session, 28th Parliament, Speech from the Throne, 1970. Also see Appendix 4.

8 See chapters 2, 3, 7, and 9 for key elements of the account of DOE/DINA interactions over offshore drilling. Also see Appendix 4.

9 Chrétien, J., *A report to the standing committee on Indian Affairs and Northern Development on the government's northern objectives, priorities and strategies for the 70's, Science and the North* (Ottawa:Information Canada, 1973).

10 See chapter 9 for a discussion of the Environmental Assessment Review Process.

11 For details of the way the system worked in the late '60s and early '70s see E.J. Dosman, *The National Interest* (Toronto: McClelland and Stewart, 1975), and a series of articles by David Crane which appeared daily in the *Toronto Star* from 11 to 18 October 1975.

12 Passmore, R.C., *Environmental hazards of northern development*, in *Transactions of the Federal — Provincial Wildlife Conference* (Ottawa: Canadian Wildlife Service, 1971).

13 Woodford, J., *The Violated Vision; The Rape of the Canadian North*. (Toronto: McClelland and Stewart, 1972).

14 Thompson, A.R., and Crommelin, M., *Canada's petroleum leasing policy; A cornucopia for whom?* (Ottawa: Canadian Arctic Resources Committee, March 1973).

15 *Oilweek, Mackenzie Delta hums with activity despite exploration and political setbacks*, 3 March 1975.

16 Sprague, J.B., *Aquatic resources in the Canadian North; Knowledge, dangers and research needs 1973, Arctic Alternatives* (Ottawa: CARC, 1973).

17 See chapter 5 for a discussion of Panarctic's reluctance to make major capital investments in the development of Arctic drilling systems.

18 See chapter 7, especially the final section.

19 Fuller, W.A., *Thoughts on the energy crisis, Northern Perspectives*, March 1973.

20 Clarke, C.H.D., *Terrestrial wildlife and northern development, Arctic Alternatives*.

Chapter 2

Drilling in Deep Water in the Beaufort Sea

In 1961 the Diefenbaker Conservative Government, anxious to give some substance to its leader's "northern vision", decided to open parts of the Arctic — including the Beaufort Sea — to petroleum exploration. There was scant geological evidence to encourage them, but the major oil companies quietly began acquiring federal exploration permits in the North and in the Beaufort Sea. What little was known about the geology of the region was promising, and there was always the encouraging knowledge that large pools of oil had been discovered in the 1940s in the US Naval Petroleum Reserve in Alaska. But in 1961, and indeed throughout most of the 1960s, the "pay-off" for Arctic exploration seemed remote. The North was inhospitable, transportation was expensive, the needed technology didn't exist, and, in an era of cheap and plentiful oil, exploration costs were astronomical. There was little enough interest in exploration on the Arctic mainland, let alone on the Beaufort Sea which was ice-covered for eight, sometimes ten, months of the year. Still, perhaps more out of interest in a speculative venture than enthusiasm for exploration, the companies steadily acquired permits giving them exclusive exploration rights on large blocks of the Sea. After all, there was little to lose. The permits were cheap, the obligations imposed on the companies were minimal, and best of all, the permits could be extended for fourteen years. Getting in on the Arctic oil play was a cheap form of insurance in case anything did turn up.

What turned up was Prudhoe Bay. In the summer of 1968 came the news that British Petroleum Ltd. and Atlantic Richfield Co. had discovered oil on the north slope of Alaska. The Prudhoe Bay discovery, 10 billion barrels of oil and 25 trillion cubic feet of natural gas, was gigantic by North American standards and big even by Mid-East standards. In Calgary, no less than in Houston and Washington, all eyes turned north. Overnight, the Mackenzie Delta, 600 miles east of Prudhoe Bay, assumed a new importance. Exploration permits that had languished for years were quickly dusted off and areas not already covered were quickly snapped up as companies realized the potential value of Arctic acreage. In the atmosphere of a latter-day gold rush, industry spokesmen confidently predicted that the western Arctic held vast quantities of petroleum, while Cabinet ministers began talking of vast pipeline networks in the North. Seismic exploration crews and drilling rigs sprouted on the Mackenzie Delta, all in search of huge reserves like Prudhoe Bay. The insurance policy had matured.

The Physical Environment of the Beaufort Sea[1]

The Beaufort Sea is one of the seven seas which encircle the Arctic Ocean. Its southern boundary is the northern edge of the continental landmass from Point Barrow at the northernmost tip of Alaska to Cape Parry at the entrance to the Amundsen Gulf on the east. Its northern boundary is arbitrarily defined as the great circle route between Point Barrow and the southern tip of Prince Patrick Island in the Arctic Archipelago. Lying directly north of the Mackenzie Delta, it was the first offshore region of the Arctic to attract the oil industry.

It is the Beaufort Sea's ice which establishes this area as the most awesome of the Arctic environments for offshore drilling. Much of the Sea is covered by the polar ice pack which circulates slowly around the Polar Basin. The polar pack normally occurs beyond the 200-metre (600 feet) depth contour in summer and so is of immediate concern to present-day drilling crews only when storms force the ice onto the shallower areas of the continental shelf.[2] In 1970 a Beaufort Sea storm showed how this could happen.

During the period from 13 to 16 September a complex storm system moved in a southeasterly direction across Mackenzie Bay in the Lower Beaufort Sea, towards the Amundsen Gulf. During the storm there were three separate occurrences of hurricane-strength winds which caused wave heights up to 25 feet and storm tides up to 12 feet. Before the storm, the polar pack was more than 100 miles north of Herschel Island, and Mackenzie Bay was completely ice-free. Within 36 hours the winds had driven a large amount of pack ice into Babbage Bight, including remnants of old multi-year floes and a number of ice islands, one of which came aground near King Point only 300 yards offshore and broke into two large fragments.[3]

Dr. E.F. Roots, Scientific Adviser at Environment Canada, is an authority on Arctic ice. At the Northern Canada Offshore Drilling Meeting, he described the nature of sea ice, which petroleum operators will face in the Beaufort Sea:

It is ordinary sea ice, not the spectacular and (in the Beaufort Sea and Archipelago) relatively rare icebergs, that will provide the workaday problems to the planner and operator of an Arctic offshore exploratory well. It is sea ice that will be responsible for the development of most of the new techniques and most of the unexpectedly high costs of offshore Arctic operations.

There is nothing very remarkable about sea ice and at first glance it should be simple stuff. But many of you here will agree that there is scarcely a substance on earth that is so intractable, so unexpectedly complicated, so deceptively passive yet irresistible, and so frustrating to the man who wants to get on with a simple job.[4]

Another speaker, K.A. Rowsell of the federal Public Works Department, described the pressure ridges which develop on pack ice and ice islands, both of which could exert very great forces on petroleum installations:

The elements of principal concern to offshore drilling operations are the pressure ridges and ice island fragments. Pressure ridges are of frequent occurrence and incidentally are the one single feature of Arctic ice about which least is known. The prominent pressure ridges as distinct from a hummocky field of ice occur when exceptional stress is exerted by wind friction or ocean currents on a continuous ice sheet of fairly uniform thickness or of such uniformity that rafting is precluded. When the ice is unable to withstand the stress, rupture occurs along a front which meanders across the ice field. Here the ice is crushed and broken into blocks of various sizes which are piled hazardously and protrude above and below the abutting ice floes.

Pressure ridges have been observed grounded in 45 feet of water suggesting that they can have greater keel depths than this. It is suspected that in the areas of interest to offshore drilling, the keels could easily attain depths of 60 feet or more.

An important feature of pressure ridges is that they frequently survive the summer melt and reappear in field ice the following winter. This remnant is much tougher and stronger than the younger pressure ridge and is reported to be the most difficult to penetrate by surface vessels. . . .

Ice island fragments which occur in considerable numbers and a variety of sizes in the Beaufort Sea and not infrequently in Mackenzie Bay represent another area of concern in offshore drilling operations. These are fragments of much larger ice islands which originate in the Ward Hunt ice shelf on the northwest coast of Ellesmere Island. From there the ice islands enter the various drift patterns in the Arctic Ocean and are continuously breaking into fragments along a crack system which appears to run in a preferred direction; probably parallel to shore in their in situ location.[5]

Few southerners have ever heard of still another kind of ice which Dr. Roots described:

The third type of ice found in Arctic waters has not received much attention and the difficulties it may introduce seem insignificant beside the massive problems of sea ice and icebergs. The suspended ice crystals are discrete separate crystals, usually less than half a millimeter in length, found in the upper few meters of water exposed to sub-zero temperatures. The crystals themselves are usually very low in salinity, but they may carry on their surfaces high concentrations of ions that alter the bulk electrical and chemical behavior of the water. Such crystals are widespread but in any one instance apparently usually transient phenomena, transitional to the formation of a sea ice crust. In areas of persistent open water they may have an important influence, because under the right conditions they could bring about rapid solidification.

Ships in Arctic waters have encountered problems at water intakes and in pumps because of rapid consolidation consequent upon agitation of suspended ice crystals, and it is conceivable that a fixed drilling platform could experience this in greater degree. However, the main importance of suspended crystals with regard to offshore operations is that they influence the behavior of pollutants and the effectiveness of pollution remedial measures.[6]

The importance of the bottom of the sea is often neglected, given the awe-inspiring nature of the ice and temperature components of the physical environment. However it too is of importance, as Dr. Roots indicated:

The rocks of interest for their possible petroleum deposits offshore are mantled with a layer of recent sediments, which are for the most part a direct continuation of the underlying sedimentary assemblage and an integral part of it, punctuated, as is most of the rest of the assemblage, by discomformities and erosion intervals as the earth surface was repeatedly above or below sea level or subjected to sea-floor scouring. These sediments are unconsolidated, and range from less than a metre to probably hundreds of metres in thickness. The upper surface of these sediments is almost everywhere gouged and scarred by passing downward-projecting pieces of floating ice.[7]

Over considerable areas of the shallow continental shelf of the Beaufort Sea these sediments are frozen, that is, there is permafrost under the sea. Jim Shearer, an expert in Arctic surficial geology, provided this description of how permafrost may have developed and changed as land areas were submerged by the encroaching sea:

The uplift of the continental shelves bordering the Beaufort Sea during the last Wisconsin glaciation resulted in the exposure of the shelf to low mean subaerial temperatures such as those which occur at the Tuktoyaktuk Peninsula at the present time. This low temperature regime existed for 10,000 or 20,000 years and resulted in the formation of a permafrost layer some 200 to 300 metres thick. Upon transgression of the sea and submergence of these previously exposed areas the regime of permafrost growth was altered. Presumably as the sea transgressed over any given point a positive temperature existed for a few thousand years when water depths were less than 15 to 20 metres. This being the case, degradation and melting of any pre-existing permafrost would have taken place during this time. With a mean annual temperature just above zero, it is thought that melting of the top down to some 50 to 70 m below the bottom would have taken place.

As the sea level continued to rise, the advent of negative temperatures (-0.5°C to -1.5°C) would have reversed the previous trend and caused renewed downward freezing. This new temperature regime would have had very little effect on slowing down the melting upwards of the bottom of the permafrost layer from geothermal heat sources. The reason for this is that the new equilibrium depth for the base of the permafrost layer (i.e. where heat loss from the surface is equal to geothermal heat gain) is much shallower than when temperatures were much lower (-10°C).

Permafrost in this sense is a relict phenomena, but is preserved by presently existing conditions. Not all areas offshore became frozen when exposed above sea level during the classical Wisconsin. As on the Tuktoyaktuk Peninsula now, large lakes where the winter ice thickness did not come into contact with the bottom have a perennial layer of unfrozen water underneath. Likewise, underneath deep river channels (six feet deep) the soils will be unfrozen. With the onset of the post-Wisconsin marine transgression and replacement of the unfrozen water (in these previously protected deep lakes) by salt water at negative temperatures, unique features developed. With the advent of a negative temperature regime in these specific locations, downward migration of the zero isotherm occurred with subsequent freezing and heaving. The formation of subsea pingos (approximately 125 discovered to date) are the most significant morphological feature which occurred after the shallow shelf areas were again covered by the sea.[8]

DINA and Industry: The Preamble to Offshore Drilling

By the end of the 1972-73 drilling season, a total of 66 wells had been drilled in the Mackenzie Delta and on Richards Island. The "majors" — Imperial, Shell, and Gulf — all had made important discoveries. But the Delta had not turned out to be the hoped-for bonanza. Of the first 66 wells drilled, four had encountered oil, 11 had discovered gas, and the remaining 51 were dry. By the end of the 1972-73 drilling season, it was clear that finding the reserves to justify building oil and gas pipelines to the south would be a long, arduous, and expensive job. Unlike Prudhoe Bay, where huge volumes of hydrocarbons were locked in a single pool, deposits in the Delta appeared to be isolated in numerous smaller pockets.

The pattern of discoveries on the Delta, however, pointed in tantalizing fashion towards the Beaufort Sea. Three of the major Delta oil discoveries — Ivik, Mayogiak, and Atkinson Point — were located on the seaward edge of the Delta, and so were a number of gas discoveries. In 1970, in a review of northern exploration prospects, *Oilweek* had pointed out that the Beaufort Sea was a very promising area:

Main offshore activity has been concentrated in the Beaufort Sea where marine seismic surveys can be conducted during the open water season over the continental shelf. All the attributes characteristic of a major oil field in other parts of the world are present, including great thicknesses of relatively young sediments.[9]

Following two more years of geophysical and seismic work, prospects in the Beaufort Sea appeared even better. In late 1972, the *Globe and Mail* "Report on Business" reported that:

oil and gas finds on the edge of the Arctic mainland ring the ocean basin from the North Slope of Alaska to the Mackenzie River Delta in Canada's North. Evidence from geophysical programs conducted in Arctic waters surrounding the islands reveals submerged structures equal to and possibly larger in size than the same favorite exploration targets found exposed on dry land . . . .

Offshore drilling is still a research project to the same industry that, in a few short years, mastered the sophisticated technology of land-based operations . . . .

Offshore exploration, regarded as the second phase of the Arctic oil and gas search, could also become a prominent feature of industrial activity there for quite another reason. There is always the possibility that land-based Arctic oil and gas prospects may take too long to materialize or, if found, fall short of commercial considerations. Then the petroleum industry would be compelled to step up its search in Arctic waters.[10]

By 1972, the industry was indeed anxious to step up its activities in the offshore areas. Not only were the prospects enticing, but offshore permit holders had mounting work obligations to meet if they were to keep their acreage. Exploration permits for the Beaufort Sea are held by many companies. They include several names familiar to those who follow petroleum exploration plays in Canada. Major permittees are Hunt International Petroleum, Dome Petroleum, and Imperial Oil; Aquitaine, Cigol, Global Marine, Gulf Oil, Hudson Bay Oil and Gas, Sunoco, and Texaco are also represented.

A consortium of 10 companies, under the name of the Beaufort Sea Task Force (BSTF),[11] has permits for approximately 1.1 million acres in a particularly promising area of the sea. The five "work bonus blocks" to be explored by the Beaufort Sea Task Force were acquired in a public tender in 1969. The terms of the permits required that the companies spend $15.5 million by January 1975 on exploration work above and beyond the normal work obligations attached to all exploratory permits.[12] The size of the tender virtually dictated that the companies would have to drill by 1975 since such a large amount of money could not be spent on geophysical exploration alone. The BSTF was formed in March 1972 to develop the drilling system needed to meet the permit obligations.[13]

It was immediately apparent that quite different technologies would be required for drilling in the shallow parts of the Beaufort Sea and the deeper waters. The Beaufort Sea Task Force, Hunt International, and Dome Petroleum, however, were all faced with developing mobile, floating drilling systems that would be capable of drilling in ice-infested waters from 150 to 1000 feet deep. The members of the BSTF were under particular pressure due to the terms of their permits. If the companies did not spend the required sum, the agreement called for the deposit money and the bonus block acreage to be forfeited. An official of the Task Force later recalled that "immediately after being awarded the permits (in 1969) the various companies began to look at ways to meet their obligations. The more the area was beheld, the more awesome the task became."[14] This awe gave way to the realization that permit holders in the Beaufort Sea had a lot to lose by pursuing separate approaches and much to gain by cooperating to share the costs of preliminary feasibility studies. Thus in 1970 the Arctic Petroleum Operators' Association (APOA) was formed because "Industry has recognized for some time that there will be many severe problems in operating in ice-infested waters."[15]

Members of APOA commissioned three engineering feasibility studies that later formed the basis of both Hunt's and the BSTF's drilling proposals. These studies, which considered the feasibility of a variety of different drilling systems, were carried out at a combined cost of $320,000 by some of the biggest names in the engineering and consulting fields: Westburne, Fenco, Foundation, Sedco, and Global Marine. When they were completed, the studies for both organizations recommended the use of ice-strengthened drilling barges for the Beaufort Sea. Convinced that such equipment was the answer, the companies who were later to form the Beaufort Sea Task Force in December 1971 made a presentation to Indian and Northern Affairs and requested concept approval, so they "could proceed with building the equipment. Indian and Northern Affairs chose not to grant approval at that time and requested more detail."[16]

In the spring of 1972 both Hunt Petroleum and the Beaufort Sea Task Force made presentations to DINA on their drilling concepts. They were told that DINA could not approve the proposals before late 1972 or early 1973. Apparently officials had concluded that approval would have to be a Cabinet decision and that it would take months to prepare documents and get the matter before Cabinet.

At the Northern Canada Offshore Drilling Meeting in December 1972, both Hunt and the BSTF presented papers on drilling systems which consisted of ice-reinforced drill barges, each of which would be supported by two work boats, two helicopters, and a fixed-wing aircraft. It was emphasized that the two APOA projects had independently recommended essentially the same drilling systems. Both companies stressed that they were anxious to obtain approval of their concepts so they could get on with construction. Hunt stated its position in blunt terms:

We have a contractual commitment to commence drilling in the Beaufort Sea during the summer of 1973, subject to our receiving government approval of our drilling system . . . . In May we were in a position to immediately proceed with final arrangements for construction of the drilling system. Shipyard space was available and we had place orders on the slower delivery items of equipment and materials. In short, we were in a position to construct the drilling system and meet our commitment. At the time of our May meeting with the proper government officials in Ottawa, we were advised by them that their requirements for a drilling system were not yet fully determined, that any new requirements would be retroactive, and that no drilling system would be approved before the end of 1972 or possible early in 1973. We, therefore, were forced to put our construction plans in a state of suspension.[17]

In April 1973 DINA circulated a restricted document, "A Position Paper on Oil and Gas Exploratory Drilling in the Offshore Regions of Canada's Arctic." Part two of the document was an "Assessment of Industry Proposal of Concept for Design and Construction, Drilling Systems for the South Beaufort Sea Summer Open Window." The assessment of both systems ended with similar declarations. In considering the BSTF system, the document stated that "All the design and engineering factors that can be considered in advance of detailed engineering design are now found to be acceptable. Assuming this drilling system concept is fabricated and commissioned in accordance with the technical and operational stipulations attached to the Application for a Drilling Authority . . . . Permission to drill can be assured."[18]

The position paper on offshore drilling was the basis for the memorandum to Cabinet, "Oil and Gas Exploratory Drilling, Offshore Northern Canada," which was submitted in June 1973 and approved 31 July 1973. An appendix to the memorandum to Cabinet contained an artist's concept of the "Components, Arctic Offshore Drilling System." But the concepts for the drilling systems submitted by Hunt Petroleum and the Beaufort Sea Task Force were abandoned soon after they were approved by Cabinet. No explanation for the change in plans has ever been provided by either the Government or the companies. It appears also that the Beaufort Sea Task Force may have been disbanded because its name no longer appears when the "work bonus permittees" are mentioned.

Dome Petroleum was represented at the Northern Canada Offshore Drilling Meeting by its vice-president and by its manager of drilling operations. Nothing was divulged about the company's plans for drilling on its offshore holdings. However, in the discussion session, the vice-president requested that the meaning of "approval in principle" be clarified, and discussion of this point dominated the session.

Dome officials wanted nothing less than an airtight guarantee that permission to drill would in fact be granted following approval in principle. For their part, DINA representatives were anxious to give the companies the assurance they wanted. In reply to Dome's queries, a DINA official offered that:

After all aspects of the proposal have been assessed and we have determined that the operation could proceed with due regard for safety and for the environment, the Oil and Minerals Division will recommend to the Minister of the Department that an Approval in Principle be granted. An Approval in Principle granted to an operator would provide reasonable assurance that an Application for a Drilling Authority would be approved for a particular well in a specified area in the interval of time specified when the system is ready to begin operations.

When this still failed to satisfy the industry, one of the senior officials in DINA went further: "Approval in Principle is approval in principle by the government of Canada and that implies therefore some pretty definite agreement that the system can and will be allowed to be used."[19] Much later, after the criticism of DINA in the COPE report, officials publicly attempted to define "approval in principle" as much less binding on the Government; in fact, however, they did honour the committment made to Dome at the Offshore Drilling Meeting in 1972.

There is little doubt that Dome's concept for an offshore drilling system was well advanced at the time of the offshore drilling meeting. According to *Oilweek*, "Dome made applications for approval of its current system in July 1973, and had the programme approved in February 1974."[20] However, official correspondence records that the application was made on 30 October 1973, and the approval in principle was granted on 1 March 1974.[21]

Although the details are not clear, it is apparent that abandonment of the barge concept and approval of drillships was the subject of active negotiation between companies, and between companies and DINA. *Oilweek* offered the explanation that "Hunt failed to get federal approval for its proposed system. Dome then decided to move in and develop its current programme. Now Hunt Oil will pay the entire cost of the first well and Dome has an option on Hunt acreage in the area."[22]

However, the timing of these moves suggests that this explanation is less plausible than one offered in a later *Oilweek* article. It suggested that the nature of the drilling system, and who would operate it, had been resolved by "negotiations" within the industry:

Dome is a substantial acreage holder offshore the Mackenzie Delta. Its net Beaufort Sea rights are 2,360,000 acres with royalty interests on another 2,907,000 acres. The company has also been able to negotiate a number of agreements which will enable it to eventually control exploration and production on 4.8 million acres in the area. This was achieved as a result of its commitment to enter into Arctic offshore drilling and the parallel commitment to develop a commercially viable drilling system for this environment.[23]

C.O. Nickle's *T.V. Oil Report* gave a description of the proposed operations which brought out that negotiations were with members of the Beaufort Sea Task Force as well as with Hunt Oil:

Canadian Marine Drilling, a wholly-owned subsidiary of Dome Petroleum, is providing the offshore drill system — one in which about $120 million is being invested. It consists of two drillships, now being built and equipped at Galveston, Texas; four supply boats and support barges, now being built on the British Columbia Coast; and related onshore facilities. The drillships are ice-reinforced, capable of drilling offshore Beaufort during the summer season. The supply boats will also be ice-reinforced, and will have ice-breaking capability to assist in support of the drillships.

The ships are expected to be completed, ready to move from southern harbours to Beaufort this coming summer. From September arrival through the winter months, various experimental operations will be carried out. When ice begins to break up in the late spring of 1976, drillships will be moved to sites of initial exploratory drilling.

Dome's Canadian Marine will anchor one of the drillships in some 200 feet of water, around seventy miles north of the Imperial-Adgo oil and gas discovery which was drilled from a temporary gravel and frozen mud island in very shallow waters immediately offshore from the Mackenzie Delta mainland. Adgo was the first Beaufort Sea offshore discovery, and first find in the Arctic world made by drilling from platforms dredged from sea-bottom material. Such islands, of course, are not feasible for deeper waters of Beaufort.

Dome's Canadian Marine drillship will test a large structure underlying its initial venture, that was outlined by seismograph on a Canadian exploration permit originally acquired several years ago by Dome Petroleum, and then farmed out to Hunt International. Cost of this well, $30,000,000 or more, will be borne by Hunt. . . .

The second Canadian Marine drillship will be positioned about 55 miles east-southeast of the Hunt venture, in about 90 feet of water, at a site some 50 miles north of the Imperial-Moyagiak oil discovery onshore on the Tuk Peninsula. Cost of this venture will be borne two-thirds by Dome Petroleum and one-third by Gulf Oil Canada. It is on Canadian exploration permits shared by Gulf, Mobil and the

French companies, Elf and Aquitaine. Dome has acquired a farmout from Mobil and the French companies. Gulf Canada, which has a record of success in discoveries on the MacKenzie Delta mainland, decided to join Dome in the offshore Beaufort venture.[24]

DINA/DOE — The Struggle for Control of Environmental Protection

The federal Government's interest in the use of the Beaufort Sea for the production of oil and gas had been centered, until late 1971, entirely within the Department of Indian and Northern Affairs. During most of the 1960s, that interest had been quietly handled by the Oil and Minerals Division of the department, the division which had been issuing exploration permits since 1961. But since the 1968 boom in northern oil exploration, the department's administrative scope and interest had expanded. In 1968, the Division of Water, Lands and Forest was formed within the Northern Economic Development Branch. This division enforced the Land-Use Regulations and regulations which were promulgated in 1972 under the Arctic Waters Pollution Prevention Act. Despite these administrative changes, intended to foster environmental protection, there was no apparent concern about the impact which petroleum activities might have on the Beaufort Sea or on native people. Had such concern existed, the department might have adopted a different position when the permits for "work bonus blocks" were issued in 1969. Since the companies were required to spend an additional $15.5 million above normal obligations, it would have been a simple matter to require that a portion of the "work bonus" funds be spent on specific environmental studies. Such work could have provided the basic knowledge for an assessment of the potential environmental impact of drilling operations and oil spills.

Certainly the need for such background research was obvious by 1969, and it was being widely discussed in the light of oil developments in Alaska. Events in Canada at that time also pointed to the dismal lack of information about possible spills in the North. The Canadian Wildlife Service, then a part of DINA, had commissioned a study on the environmental effects of oil pollution which reported, in mid-1969, that almost nothing was known about the effect of an oil spill in the Arctic, or about methods of containing or cleaning up oil spilled in Arctic waters.[25] The same year, conservation problems and research needs in the Arctic were highlighted at an international tundra conference at the University of Alberta at which the chief of the DINA Water, Lands and Forest Division presented a paper entitled "Conservation in Canada's North."[26] About the same time, the DINA Minister, Jean Chrétien, began to sprinkle his speeches more and more liberally with the theme of environmental protection. Nevertheless, little was done to consider the specific need for research on the impact of petroleum activities on the Beaufort Sea. Neither DINA, nor the Department of the Environment (DOE), nor its predecessor, the Department of Fisheries, had seriously considered the environmental issues when the oil companies began to press for approval to develop offshore drilling systems for the Beaufort Sea.

The Department of the Environment was first brought into the negotiations in June 1972, at the meeting with the Beaufort Sea Task Force. Both DINA and industry officials, however, seemed less interested in any expertise the department might have to offer than in whether DOE was likely to take any action to delay the start of drilling. At that meeting, DOE was specifically requested to make its position clear to DINA, in order that a position paper could be prepared for the Minister of Indian and Northern Affairs before the end of 1972.

Early in September, DOE formed an ad hoc working group with representatives from the various services in the department to evaluate whatever information existed and to define the additional research needed to assess the potential impact of offshore operations. But the committee floundered and DOE lost an important opportunity to make a strong case for environmental protection at the Northern Canada Offshore Drilling Meeting. Had it vigorously presented its case then the course of subsequent events might have been much different.

Despite Hunt Petroleum's unhappiness with DINA's performance, it also became clear at the Offshore Drilling Meeting that the interests of DINA and of industry were converging and that both were wary of potential opposition from DOE. In his opening speech to the meeting, the Assistant Deputy Minister for Northern Affairs, Digby Hunt, subtly defined the department's stance:[27, 28]

People are asking why we need to know about the resources of the Beaufort Sea at this time. They are asking why do we need to explore for and exploit the resources of the North; what is our hurry? Surely there are enough fossil fuels, not only in southern Canada, but elsewhere in the world, that we don't need to take into the North this tremendously disrupting force, this new technology that has such a high impact on the environment and on the local people. Why do we not go a little slower; why don't we look elsewhere? In other words, the development ethic is being challenged and we in government and you in industry ignore that challenge at our peril. It is a very real challenge and we must meet it; we must answer those questions . . . . So, while this meeting here does not have as its purpose those questions, I did want to bring them to your attention because, as I have said, they are not going to go away. Unless you can provide persuasive answers along with us, those who advocate the other approach are going to gain credibility.

One suspects that most of the conference attendees slipped out for coffee during the DOE papers, which were, for the most part, vague and offered little insight into the kinds of research needed or the time required to conduct research programmes. Only one DOE scientist, near the end of the conference, made a strong case:

I am almost taken aback at the amount of work that has been conducted in the region with respect to feasibility studies, and the evaluations, primarily of physical environmental factors, as they influence engineering feasibility.

I feel that there has been relatively little emphasis placed on the biotic community and the ecological implications of Arctic development. I feel that in view of this lack of emphasis . . . that there may be misinterpretation such that people get the feeling that we know enough about the biotic communities. I would like to stress that at present we are in the infant stages of determining the complex relationships of the Beaufort Sea and the Mackenzie Delta, and we have a very meagre understanding of the interrelationships of living things in these areas. For this reason I would simply like to make the point that there is a lot of work to be done not only finding out what is there biotically, but why it is there and what influences man has on it.[29]

The pace of events quickened during 1973. Early in January, an interdepartmental meeting on the Beaufort Sea was held in Edmonton. In Ottawa, the DOE ad hoc working group continued its attempts to formulate a programme of environmental research for the Beaufort Sea. But DINA had no intention of leaving the initiative in environmental matters to DOE. It had dealt successfully with opposition from DOE on land-based operations by setting up the Land Use Advisory Committee. DOE had three seats on the committee, but DINA maintained firm control through the chairman and by insisting that the committee's function was strictly advisory.[30]

In February DINA formed an ad hoc, temporary version of the Arctic Waters Oil and Gas Advisory Committee (AWOGAC). Its purpose was to assist in researching environmental aspects of offshore drilling in the Arctic for the memorandum DINA was preparing for its Minister to submit to Cabinet. This committee was to be followed by "A similar interdepartmental committee in Yellowknife, parallel to the Land Use Advisory Committee, [to] deal with the environmental aspects of individual wells when operations commence."[31] In terms of interdepartmental politics the role of the ad hoc committee and its successor was to help DINA head off any move by DOE to obtain regulatory powers over environmental aspects of offshore drilling through the Fisheries Act or the Canada Water Act. To make certain that DINA had the same degree of control over offshore operations as it had over those which were land-based, it presented the memorandum to Cabinet on offshore drilling in which "The Minister . . . recommended that the Government approve in principle the exploration for and development of the potential oil and gas resources in the offshore regions of Canada's Arctic by reconfirming the authority of the undersigned to approve and license such activities."[32]

The extent of the role played by the DOE representatives on the Committee is not clear; at any rate the draft memorandum, which was sent to DOE for information on 30 May, seriously understated the hazards and risks involved in offshore drilling. It referred to favourable precedents in southern Alaska, the cold northern waters of the North Sea, and the Bass Straits of Australia, but failed to mention the near-disaster at the first offshore operation in Hudson Bay[33] — a Canadian precedent more relevant than any of the three mentioned. The memo summed up the environmental risks in this paragraph: "Although the inherent risks of accidental pollution of the Arctic marine environment cannot be reduced to zero, the risks are considered to be low and reasonable with respect to probable national and regional economic and social benefits to be derived."[34]

The Cabinet memorandum was reviewed by DOE's Management Committee on 21 June. The Committee was advised that the departmental working group considered DINA's analysis of the environmental aspects of offshore drilling inadequate and the minutes indicated that DOE members considered that their department should be more directly involved in the regulation of the offshore drilling:

The (Management) Committee agreed that the Arctic is a critical area, environmental conditions are extreme and the likelihood of a spill is significant. Environmental conditions under which licenses are issued must therefore be set jointly by DINA and DOE.

The consensus was that the Department should react to DINA's request for information, and our reaction must be based on a proper understanding of the role played by DOE in the preparation of the submission. A recommendation should be made that the wording is too absolute to correspond to our views; although DOE may agree with the unfortunate but necessary decision to proceed with oil drilling, such activity should be carried out under a set of guidelines which will provide safety measures and the accumulation of data for future guidelines of operations. Members agreed that the area of responsibility and role of DOE needs to be clearly established through negotiations. Full understanding between Departments on matters of principle and of application and field operations (i.e. monitoring on an ongoing basis) is needed.

DECISION (23.7) It was decided that:

(a) *on the basis of the above comments, the ad hoc working group . . . will review the DINA Memorandum to Cabinet and provide critical comments to the Management Committee. These comments will be used to brief the Minister, and they will be forwarded to DINA if deemed appropriate.*[35]

There is a gap in the documentary evidence of DINA/DOE infighting following the meeting of DOE's Management Committee. At any rate, on 31 July 1973, the Cabinet gave approval in principle to the exploration for and development of offshore oil and gas resources and reconfirmed the authority of the Minister of Indian and Northern Affairs to approve and license offshore operations. The Cabinet decision apparently included a proviso that all projects under federal jurisdiction comply with policies and requirements of other government departments.

Subsequent actions taken by DINA, with little regard for the concerns of DOE, in authorizing drilling from artifical islands and ice islands, and in giving approval in principle to drilling in Lancaster Sound, give clear evidence that DINA was successful in gaining ascendancy over DOE in offshore operations, just as it had in those which were land-based.[36] The principal concession made to DOE was approval of a crash programme of baseline and environmental impact studies to be carried out before drilling began in deepwater areas of the Beaufort Sea. The studies were to be paid for by the petroleum industry. The research project and the political aspects of the Beaufort Sea Project are reviewed in Chapter 9.

The Native Peoples' Concerns and Interests

On 8 February 1974, COPE (The Committee for Original Peoples' Entitlement)[37] released its report on offshore drilling in the Beaufort Sea.[38] On the previous weekend COPE's Board of Directors had reviewed the report at a meeting in Paulatuk. It was the first concrete information that native people had obtained on industry's plans to undertake a major programme of offshore drilling in the Arctic. In 1971 and early in 1973 the Tuktoyaktuk Hamlet Council had reluctantly agreed to the construction of the first two artificial islands, Immerk and Adgo, but these had been presented as simple extensions of land-based operations. The Inuit people of the region had not been given any information either on Imperial Oil's plans to build and drill from a succession of artificial islands or on plans to drill in deep-water areas of the Beaufort Sea.

The Board of Directors also considered and approved the wording of the news release which was issued a few days later from COPE's office at Inuvik.[39] The release reviewed the salient points made in the report and stated that it appeared "that balanced, long-term development . . . is being sacrificed for immediate profit and a panic reaction to the energy crisis . . . . It is appalling that neither COPE, Inuit Tapirisat of Canada [ITC], Settlement Councils, nor hunters' and trappers' associations in the region have been consulted." It stated that the Board of Directors had requested the federal Government to begin immediate consultation with COPE and ITC on the whole matter of offshore drilling in Arctic waters.

The report and the news release were widely circulated. Virtually every major newspaper in the country carried a story under banner headlines: "Ottawa accused of backing hazardous oil drilling" (*Edmonton Journal*), "Native group warns of oil spills in the Arctic" (*Toronto Star*). The story was also featured on national news radio and television and on CBC radio's popular show, "As it Happens."

COPE had sent advance copies of the report to DINA and DOE, so Digby Hunt of DINA was ready with the Department's official reply when he was interviewed by Canadian Press. He focused his remarks on the approval in principle aspect of offshore drilling. The C.P. story stated his case: "Hunt said approval in principle was given to an exploration company based on its technical plans. But he said the Government also wanted to be satisfied that no lasting environmental effects would result. Until then no drilling would be approved. Hunt also declared that full consultation with native people would precede drilling. Replying to criticism by the Inuvik group he said "there is nothing to consult about now since there are no firm plans."[40] On 6 March, less than a month later, Chrétien made an announcement about the Beaufort Sea research programme. The last paragraph of the communique clarified what Digby Hunt had meant by full consultation with native people and about there being no firm plans for offshore drilling: "Dome Petroleum Limited and Hunt International Petroleum, two of the companies with permits in the region, have been authorized to commence construction of drilling systems that will be employed in the Beaufort Sea. Mr. Chrétien explained that this would allow the companies to move the specially designed systems to the Beaufort Sea region probably late in 1975 so that they would be ready to commence drilling operations at the beginning of the open water season of 1976."[41]

COPE's investigation of offshore drilling began early in October 1973. Shortly after joining COPE's staff as a resource worker, I attended an environmental conference held in Yellowknife under the sponsorship of the Arctic Petroleum Operators' Association (APOA). Offshore drilling was not on the agenda, but an immediate enquiry to Bill Armstrong (then Regional Director of DINA in the NWT) confirmed that plans were to commence drilling in 1975. On returning to Inuvik, COPE's President, Sam Raddi, instructed me to conduct an immediate investigation and to report to COPE as soon as possible. On 1 November, I wrote Mr. Chrétien requesting specific information on studies to be conducted and copies of any public information. The letter was duly acknowledged by an assistant, but by the time I received Mr. Chrétien's reply on 29 January the report had already been written and submitted to COPE. The Minister's letter clearly stated the department's reluctance to part with any information:

I am sorry, but we are unable at this time to provide you with the information you require. While plans are, as you say, being made for studies to investigate the environmental aspects of exploratory drilling for oil and gas offshore in the Beaufort Sea, the outline of these studies has not yet been finalized nor has the division of responsibilities between Government and industry. I might add that I have made no public statements about these studies nor, to the best of my knowledge, have there been news releases or talks by any of my officers on the subject.

We will, of course, be pleased to give you the information you want as soon as it is available, but I cannot as yet say when this will be.[42]

In the meantime, I learned that 180 industry and government officials had met at a Northern Canada Offshore Drilling Meeting held in Ottawa in December 1972. Mr. Armstrong told me that DINA's head office had advised him that copies of the proceedings would not be made available, except to those attending the meeting. The matter was pursued from Ottawa by CARC. As in the case of my letter to Mr. Chrétien, delay appeared to be an important tactic. A letter to Mr. Hunt on 31 October resulted in a reply from a subordinate a month later and a final letter on 18 January. The end result was that we received copies of the papers presented by members of government and an edited version of the APOA paper on research, but neither the papers presented by industry nor the discussion section of the proceedings. The explanation was difficult to accept: "The industry papers contained in the proceedings cannot be released. It was only with the understanding that copies of papers presented by industry would be made available to participants invited to attend the meeting, that the Department of Indian Affairs and Northern Development was able to ensure meaningful industry participation and discussion on relevant matters."[43]

By the end of November it was clear that information on offshore drilling plans and proposals would not be available from official sources. At this point we changed our approach and searched out unofficial ones. Research for the COPE report and a considerable portion of this book was based largely on sources of information which are not normally available outside government and industry.

After the release of the COPE report, Sam Raddi sent copies of it and of the news release to all provincial and national native organizations, to the majority of scientific societies, to major conservation and environmental organizations in Canada and the US, and to a number of international organizations including the International Union for the Conservation of Nature and Natural Resources (IUCN) in Morges, Switzerland. An abridged version of the report was also sent to every individual member of COPE.[44] A special effort was made to have the topic of offshore drilling featured on local and regional radio programmes. Within a month of the release of the report there were few individuals in Delta or Beaufort Sea communities who were not aware that the oil companies were going to sea in their search for oil and gas. The Inuit throughout the Canadian Arctic were also informed through *Inuit Monthly,*[45] a publication of the Inuit Tapirisat of Canada.

COPE kept up its interest in offshore drilling. Many letters were written to the Ministers of the Environment and of Indian and Northern Affairs to obtain additional information or clarification of policy; assistance was provided members of trappers' associations and settlement councils who were reviewing land-use applications for the construction of new islands; new information on developments was passed on through the Directors, newsletters, radio, or memos to key individuals. However after the COPE report was made public much more information began to be available from government and industry on offshore drilling. In the late spring, DINA's Regional Information Services issued an offshore edition of *Dialogue North.*[46] In the fall, APOA sponsored the publication of the *Beaufort Seer,* to provide information on the Beaufort Sea environmental program;[47] it also conducted, through Pallister Resources Management Ltd., two series of community meetings in October and January and released a film, "Understanding the Beaufort Sea."

The Government and APOA publications and the film presented offshore drilling in positive, public-relations terms which were completely acceptable to the two proponents. The *Beaufort Seer* concentrates on descriptions of the various aspects of the research project and on accounts of the community meetings. There is no editorializing on the hazards and risks of offshore drilling or on the way development could impinge on the interests or traditional rights of the native people.

COPE continued as an independent watchdog despite the combined efforts of government and industry. For example, the January issue of the *Beaufort Seer* reported on studies of the impact of oil on ice which were being conducted under controlled conditions in Balaena Bay near Cape Parry.[48] However, no mention was made either in the story or in the community meetings of plans to release a quantity of oil under ice in an offshore area. Wild rumours about the amount of oil to be released circulated in Beaufort Sea communities. COPE became aware of the situation and sent a telex to the Ministers of DINA and DOE on 25 March:

Cope understands that the Department of Environment plans to conduct an experiment involving the discharge of crude oil under the ice. These tests are to take place in early April in an area offshore of Cape Parry, N.W.T. Neither COPE nor the Hamlet councils of Tuktoyaktuk and Paulatuk have been either consulted or officially informed of these tests — which are separate from the tests being carried on in Balaena Bay. DOE has known about these tests for a long time. In response to a query from us, Mr. Allen Milne, project manager for the Beaufort Sea project, sent us a brief outline of the experiment which was dated November 25/74. We received it March 18/75.

COPE recognizes the need for scientific experiments to determine what would happen if there was a spill or a blow out in the Beaufort Sea. However, we also feel that such experiments, particularly those which are potentially hazardous to the resources upon which we depend, should be subject to consultation and approval by us before they are allowed to proceed. We do not now have enough information about the proposed experiments to assess them, but the period between November and now would surely have been adequate.

Neither Minister responded to it nor to a follow-up telex in early April. The incident ended 17 April with a final telex from Sam Raddi to the two Ministers: "COPE understands that the offshore oil release experiments took place off Cape Parry on April 9. The original plans were substantially cut and 360 gallons of oil were released. We realize that the amount of oil finally released was very small. However, the principle that the people who live in the western Arctic should be consulted about all proposed land and water use is still valid. COPE still objects to these experiments having taken place without opportunity for review by us."

Similarly, the *Beaufort Seer* did not mention the proposal by Canmar Drilling that the Government dredge Tuktoyaktuk Harbour so the drillships could overwinter there. COPE obtained a copy of a proposal directly from the company after it had learned of the proposed project through other sources.

Nothing approaching consultations of the kind that COPE advocated in its news release has happened. The community meetings have included presentations by members of government and industry, questions and comments from the audience, but no public participation in the sense that the people involved might possibly influence decisions. According to members of COPE who attended, there has been a tendency for industry representatives to use the community meetings to try to gain acceptance of their ideas on petroleum development. Some of the research results also have been subject to misinterpretation. For example, the fact that six seals survived for 24 hours in an experimental tank where crude oil had been spilled resulted in telling native people that seals would not be harmed by oil spills (see Chapter 9).

However, such meetings undoubtedly represent consultation of the kind Mr. Hunt promised when he reacted to the COPE report in February 1973. There is other evidence which supports the view that nothing has really changed in the consultative process. For example, even the normal process of seeking approval from settlement councils for land-use activities appears to have been abandoned in the case of deepwater drilling. No community on the Beaufort Sea was consulted about the setting of caissons at the offshore well sites for drilling in 1976.

Because the operation meant drilling into the sea floor, it should have been considered by the councils. But perhaps the Board of Directors and the members of COPE can be excused if they are cynical about the future of the Beaufort Sea.

They learned that despite the smooth assurances of the government and industry public relations campaign, Canmar Drilling (a wholly-owned subsidiary of Dome Petroleum) failed in its effort to sink the two mammoth silos into the sea floor in preparation for next summer's drilling. The short summer season prevented Canmar from even commencing operations at the prime drilling location in 150 feet of water. At the second location, in 90 feet of water, the 70-foot silo was only partly lowered into the sea floor before operations were halted due to freeze-up. The top of the 30-foot diameter silo now sits only 60 feet below the sea surface. Among scientists who have worked in this area, the heavy betting is that it will not survive the winter's onslaught of ice.

References and Notes

1 An article on the biology of the Beaufort Sea is found in Appendix 4.

2 At the Northern Canada Offshore Drilling Meeting (N.C.O.D.M.) Mr. R.B. Thornborg of Global Marine Inc. stated that "We have given at a recent date some consideration to actually operating or trying to determine the feasibility of operating on the polar ice pack." See *Global Marine's Drilling System for the Arctic Offshore*, Proceedings of Northern Canada Offshore Drilling Meeting, DINA (Ottawa: December 1972).

3 Burns, B.M., *The Climate of the Mackenzie Valley*, vol. 2, *Beaufort Sea*, Wilson, H.P., *Study of the Beaufort Sea Storm of September 1970*, 1971 (unpublished)

4 Roots, E.G., *Sea ice and ice bergs* (Ottawa: N.C.O.D.M., 1972).

5 Rowsell, K.A., *Notes on some major technical problems* (Ottawa: N.C.O.D.M., 1972).

6 Roots, see n. 4.

7 Roots, *The sea floor below* (Ottawa: N.C.O.D.M., 1972).

8 Shearer, J., *The physical environment of the Beaufort Sea* (Ottawa: Northern Assessment Group, 1975), unpublished.

9 *Oilweek, GSC uncovers Beaufort secrets*, 16 November 1970

10 *Globe and Mail, Oil industry expected to intensify Arctic undersea activity*, 27 December 1972.

11 Members of the Beaufort Sea Task Force were AMOCO, Aquitaine, Canadian Superior, Elf Oil, Gulf Oil, Hudson's Bay Oil and Gas, Mobil Oil, Texaco Exploration, and Union Oil. Imperial Oil joined later and became the tenth member of the group.

12 Canada, DINA, Oil and Minerals Division, *A position paper on oil and gas exploratory drilling in the offshore regions of Arctic Canada* (Ottawa: DINA, 1973), restricted.

13 Swift, U.N., *An exploratory drilling system for the Canadian Beaufort Sea* (Ottawa: DINA, 1972).

14 Swift, see n. 13.

15 Wilkins, F.L., *Arctic offshore knowledge and research — Industry* (Ottawa: DINA, 1972).

16 Swift, see n. 13.

17 Osborne, J.C., Sanford, R.M., and Haynes, J.O., *A proposal for offshore drilling in the Beaufort Sea* (Ottawa: DINA, 1972).

18 Oil and Minerals Division, see n. 12.

19 Dunkley, C.S., *Record of discussion, question and answer period* (Ottawa: DINA, 1972).

20 *Oilweek, Dome bares five-year Beaufort programme to drill four $20 million holes yearly,* 8 July 1974, and box on Dome-Hunt arrangements.

21 Letters from A.D. Hunt, Assistant Deputy Minister (March 1974) and the Hon. Jean Chrétien, Minister of DINA (3 April 1974) to Mr. J.P. Gallagher, President, Dome Petroleum.

22 *Oilweek,* box on Dome-Hunt arrangement, 8 July 1974.

23 *Oilweek, Beaufort drilling details unveiled with growing Dome fleet and concept,* 7 October 1974.

24 *Daily Oil Bulletin, Beaufort Sea drilling planned for 1976 will likely be the costliest offshore wells ever drilled,* 3 February 1975.

25 Pimlott, D., Kerswell, J., and Bider, R.J., editors, background study for the Science Council, Special Study No. 15, *Influence of Resource Development on Fisheries and Wildlife* (Ottawa: 1971).

26 Naysmith, J.K., *Conservation in Canada's North,* in *Productivity and conservation in northern circumpolar Lands,* edited by W.A. Fuller and P. Kevan (Morges: I.U.C.N., 1970).

27 Hunt, A.D., *Welcoming address and opening remarks,* to the Northern Canada Offshore Drilling Meeting, December 1972.

28 During 1972, DOE was developing an organizational structure after having been reconstituted from the Department of Fisheries in 1971. There was much concern about the conflict over the formation of new units, such as the Environmental Protection Service, which were assuming some of the responsibilities of established units. It is possible that internal conflicts of this nature were an important factor in DOE's inability to deal effectively with environmental aspects of offshore proposals made by DINA.

29 Surrendi, D.C., at N.C.O.D.M., December 1972.

30 Usher, Peter J., and Beakhust, Grahame, *Land regulations in the Canadian North, Northern Perspectives,* November 1973.

31 Oil and Minerals Division, see n. 12.

32 DINA, Jean Chrétien, *Oil and gas exploratory drilling offshore Northern Canada,* Draft Memorandum to Cabinet, May 1973. See Appendix 1.

33 See Chapter 6, "Drilling in Hudson Bay."

34 Chrétien, see n. 32.

35 DOE, Minutes, DOE Management Committee, 21 June 1973.

36 During the period from mid-1973 through 1974 there was a flow of internal documents in DOE dealing with developing an effective DOE/DINA interface. It was evident from these that much of the pressure for DOE to exercise its legislative and regulatory prerogatives came from operating levels and received little support from senior administrators or from the Minister. The ruling to staff that DOE would play a supporting role to DINA appears to have developed from a meeting held on 29 October 1974 between senior administrators of DINA and DOE. The first paragraph of the memo, as reported by Jean Lupien, an Assistant Deputy Minister, stated: "The first item reviewed was the establishment of programs by DOE. We recognized that DINA was a major user of baseline data on northern ecology and that they had as well the responsibility to coordinate all federal activities in the North. In addition, they have specific priorities about which we can be an essential contributor. For these reasons, we agreed that mechanisms should be established to ensure that DOE's involvement in the North should be discussed at the appropriate time with DINA to ensure that the interests of both Departments are met."

Action against the Giant Yellowknife Mine was one of the rare examples where DOE asserted itself and pressed charges without DINA's acquiescence. The case was discussed at the meeting and resolved in a way that indicates that such freedom of action will not be repeated in the future. DOE will be the specialist and DINA the regulator where pollution incidents occur. This is the way the memo stated the case: "The Giant Yellowknife Mine's requiring legal action by DOE is raised by DINA to provide a better understanding as to under what statute or authority DOE would act in the case of a spill of toxic material. I have undertaken to raise with EPS that we should be considered by DINA as the federal specialist through our emergency centre in the case of spills of any toxic material whether it takes place on land or in water. As such, we would immediately be called upon the scene of an accident and give our best advice to DINA. DINA would act on such advice recognizing that the issuer of the permit should be involved directly in any instruction to stop activities. In cases where a legal action should be taken, DINA would join DOE in such legal procedure."

Above all, DOE must not infringe on DINA's territory by checking for infractions of the Fisheries Act where DINA has responsibility under one of its acts: "That discussion applied also in the case of the spill from Gulf Oil in the Yukon. DOE in addition undertook to examine with EPS the need for EPS to stop routine inspections which are carried out and are part of the responsibility of DINA."

The message conveyed by the Memorandum was very clear. DINA had gained complete ascendancy over DOE in the N.W.T. and the Yukon in environmental protection. The spirit of free-wheeling development would now have clear sailing and would not have to worry about swatting DOE mosquitoes.

37 The Committee for Original Peoples' Entitlement is a regional native association in the western Arctic. It is affiliated with Inuit Tapirisat of Canada but represents the interests of both Indian and Inuit people of the region.

38 Pimlott, D.H., *Offshore drilling in the Beaufort Sea* (COPE, January 1974), mimeographed.

39 COPE, *Drilling for oil and gas in the Beaufort Sea*, press release, 8 February 1974.

40 Canadian Press, telex of story on DINA's reaction to the COPE report, made available through the CBC.

41 DINA, *Beaufort Sea — No drilling until 1976*, communique, 6 March 1974.

42 Letter from the Hon. Jean Chrétien to Douglas Pimlott, 29 January 1974.

43 Letter from Mr. Fred Joyce, Director, Northern Natural Resources and Environment Branch to Mr. K. Vincent, CARC, 28 November 1973.

44 Pimlott, D.H., *A summary report on offshore drilling in the Beaufort Sea* (COPE, January 1974), mimeographed.

45 *Inuit Monthly, Offshore drilling in the Beaufort Sea*, February 1974.

46 March 1974. In mid-August 1974, a modified version of the offshore issue of *Dialogue North* was circulated to some members of the media and to some conservation organizations. These paragraphs occurred on the first page:

Dialogue North owes a debt of gratitude to COPE (The Committee for the Original Peoples' Entitlement) for making this issue on offshore drilling possible. COPE's news release, and its report on Offshore Drilling in the Beaufort Sea were released in early February. The subject had been Top Secret until then; however, the COPE release made further secrecy futile, so an immediate decision was made to publish this issue.

Dialogue North regrets that it was not possible to include a section on what will happen when oil spills occur in the Beaufort Sea or in the Arctic Basin. That subject is a delicate one and has not yet been cleared for public discussion at official levels. It is anticipated that an Oil Spill Edition will be published after the first major spill occurs in Arctic offshore waters.

A story carried on CP wire services about the matter appeared in the *Globe and Mail* on 20 August. The statement was repudiated by DINA who branded the insertion as false (*Globe and Mail*, 23 August). The revised edition was also distributed from Yellowknife.

47 *Beaufort Seer*, Pallister Resource Management Ltd., 3rd Floor Pacific Plaza, 700 6th Avenue S.W., Calgary, Alberta T2P 0T8.

48 *Beaufort Seer, Oil impact on ice part of studies*, January 1975; see n. 47.

Chapter 3

Drilling in Shallow Water

Immerk, the first artificial island constructed and the first wildcat well to be drilled in the Canadian Arctic Ocean, set an important and unfortunate precedent. The construction of the island was approved by DINA under the Territorial Land Use Regulations early in 1972.[1] At the time, the Department of the Environment (DOE) was being organized and was barely represented north of 60; it was not in a strong position to fight for its rights to regulate offshore drilling operations even had the spirit of fight existed. But still it is interesting to conjecture how the course of events might have been altered had the situation been different. Would the subsequent story have been different had DOE had a forceful person in charge at Yellowknife? if he had fought to regulate offshore operations under the Fisheries Act, and been strongly supported by Ottawa? In fact, it might have made a difference, because *Hansard* of 1970 and 1971 was embroidered with ringing phrases about environmental protection and the problems associated with oil spills. A skillful, determined Minister in the Cabinet of 1971 might have set a different course.

But Immerk became a precedent, and later it proved impossible for DOE to make even a small dent in DINA's jurisdictional armour. By failing to challenge DINA over Immerk, DOE lost the war before it really started. When, during the summer of 1973, the District Manager of DOE's Environmental Protection Service attempted to have Adgo, Imperial's second island, established as an offshore operation under DOE's environmental mandate, it was too late. The islands were classified as land operations and were regulated by the Territorial Lands Act, a decision soon to be fortified by the Cabinet's reconfirmation of DINA's authority to approve and license offshore activities.[2]

In the meantime, DINA had formed the permanent Arctic Waters Oil and Gas Advisory Committee and had referred to it in the Cabinet memorandum. Its terms of reference were to "advise DINA with regard to the terms and conditions to be placed on drilling operations in respect to the deposit of waste and the protection of the environment." DOE had only one representative on the committee. The Adgo application was referred to the committee, but it is difficult to understand why DINA made this move. David Gee of DINA was chairman of both the Land Use and the Arctic Waters committees and so was in a position to order things to DINA's best advantage. Perhaps the move was to demonstrate to DOE that DINA had control of environmental aspects of offshore as well as of land drilling.

Five days prior to Cabinet action on offshore drilling DINA called a meeting of the Arctic Waters committee to consider Imperial's applications to construct and drill from Adgo. It was called on short notice and only two members attended. Nine days after Cabinet action DINA approved the applications, even though DOE was still in the process of reviewing environmental aspects of the application. The result was that the Adgo application was passed by DINA without even perfunctory surveillance by the Land Use Committee, and without consideration of any recommendations from DOE.

Subsequent events indicate that at that point DOE lost the chance it had to play a major role in placing environmental controls on offshore drilling in Arctic waters. From then on the mandate to review applications for artificial islands was returned to the Land Use Advisory Committee, where DOE would make recommendations and DINA would make decisions. DOE never seriously questioned the arrangement again and, as is pointed out in several chapters of this book, has never been able to do more than play a minor role as scientific advisor to DINA in other areas where offshore drilling operations are being, or are to be, conducted.

The Physical Environment of Mackenzie Bay

In the vicinity of the Mackenzie Delta, the Beaufort Sea is shallow and the bottom is overlain by deep deposits of unconsolidated sediments. These range from less than a metre to probably hundreds of metres in thickness. In many parts of Mackenzie Bay the water is no more than 10 feet deep at distances of as much as 10 miles from land, and in many areas the 60-foot depth contour is 20 to 30 miles from the mainland.[3]

The 60-foot contour is of considerable significance to offshore drilling because it represents the maximum extent to which shorefast ice normally forms in winter. Shorefast ice forms quite early to the 30-foot contour but much more slowly in water from 30 to 60 feet deep. The inner zone is usually comprised of relatively smooth ice which is fairly uniform in thickness. It provides quite a good surface for transportation and other work associated with exploration for oil and gas. The peripheral area of the shorefast ice is a transition zone where the ice surface is usually quite rough and broken. It results from the formation of pressure ridges during the period when the annual ice is forming. The ice here varies greatly in thickness. It is thinnest in recently-formed leads and thickest in the vicinity of pressure ridges, where the ice keels often extend far below the surface of the water. The sediments of the near-shore region are deeply scoured by rafted ice, pressure ridges, and ice islands which are driven inshore by the currents and winds.[4]

At the Northern Canada Offshore Drilling Meeting, K.A. Rowsell gave a vivid description of the scene:

Included in the winter ice canopy of a typical near shore region in the Arctic is ice of many kinds, shapes, strengths and thicknesses. In general this canopy is continuously changing in response to the variety of forces being applied to it. Hummocky fields are being formed, rafting is going on where the ice sheet is relatively thin, pressure ridges are being pushed up where abutting ice floes are being forced together and the ice is unable to resist the stress. Interspersed throughout this sea ice scape in the Beaufort Sea and Mackenzie Bay are remnants of old pressure ridges which survived the previous summer melt and ice island fragments from the larger ice islands in the Arctic Ocean.[5]

Present and Future Operations in Shallow Water

Imperial Oil Ltd. holds permits for a large portion of the nearshore areas of the Beaufort Sea. The company's holdings extend along the face of the Mackenzie Delta and the Tuktoyaktuk Peninsula for approximately 200 miles.[6] Its acreage normally extends to about the 60-foot depth contour, and approximately 25% of the total acreage covers waters which are less than 10 feet deep.

Imperial's strategy, described at the Offshore Drilling Meeting, was "to move seaward in a stepwise fashion learning more about the geology, the engineering and the environment with each increment of water depth."[7] In 1969, Imperial engaged consultants to recommend drilling and production methods for waters up to 60 feet deep. It was on the basis of these studies that a decision was made to construct unretained artificial islands in shallow water, particularly in water less than 10 feet deep. Since 1972, Imperial has constructed and drilled from six artificial islands: Immerk B-48 (water depth 6.5 feet), Adgo F-28 (w.d. 6 feet), Pullen E-17 (w.d. 4 feet), Netserk O-33 (w.d. 9 feet), Netserk North E-30 (w.d. 23 feet), Adgo P-25 (w.d. 6 feet). During 1975-76 it will add Adgo K-18 (w.d. 7 feet), Arnak L-30 (w.d. 17 feet), Ernerk I-21 (w.d. 22 feet), and Isserk (w.d. unknown). It will also commence construction of Issigak A-57 (w.d. 42 feet) but will not complete it until 1977; drilling will be done in 1978. Sun Oil has built Unark L-24 (w.d. 6 feet) and Pelly B-35 (w.d. 6 feet).[8] Both companies have drilled other wells from land locations in the area. Imperial, for example, drilled on Hooper Island and will drill Garry P-04 on Garry Island during the winter of 1975-76.

Immerk was begun in 1972 and completed in 1973. Drilling began on 17 September 1973 and was terminated in 1974 because Imperial encountered abnormally high formation pressures that precluded safe drilling.[9] Adgo F-28 was partially constructed during the spring of 1973 and completed later that fall; drilling began early in January 1974. The discovery of gas late that month at approximately the 7000-foot level appeared to spur interest in drilling in the shallow waters of the Beaufort. Shortly after the announcement, Imperial announced construction of the Pullen island, and Sun Oil revealed its plans for drilling from man-made islands. The subsequent announcements that Adgo had also encountered oil at three levels of the test well, and that Shell had discovered commercial quantities of oil and gas at Niglintgak M-19 on the edge of the Beaufort Sea, have served to maintain exploratory activity in the Beaufort region in the face of some political setbacks.[10] The oil strike at Adgo is the first offshore oil strike in the Arctic and could well signal the "elephant-sized" fields that the industry is hunting for. The Adgo well was drilled to a depth of 10,528 feet and was abandoned in early March 1974.[11]

A review of construction methods used for the first three islands built by Imperial — Immerk, Adgo, and Pullen — and Pelly, the second island built by Sunoco, reveals that island-building techniques have been constantly evolving. Immerk, the first, was constructed from gravel taken from the bottom of the bay by a hydraulic suction dredge about a mile from the construction site and transported to the site by a pipeline mounted on piles. Construction was supposed to require 400,000 cubic yards of gravel, and 10,000 cubic yards were to have been deposited each day. Several problems were encountered including bad weather, invasion of the area by ice, and coarse gravel which reduced the efficiency of the dredge and pipeline by more than 50%; the design of the island was modified so that only 250,000 cubic yards of gravel finally were used.[12]

When constructed, the island had a steeper slope than planned and the erosion beach was eliminated, which resulted in heavy erosion on the north side of the island during the winter of 1972-73. After the modified version of the island was completed to control erosion the slopes were covered by sheets of plastic filter paper covered by wire. These were anchored to timbers or steel pipes, and the entire surface overlain with anti-torpedo nets, surplus from World War II. The total cost of building and drilling from Immerk was about $10 million.

The concept for the construction of Adgo F-28 was quite different, since much of the construction material was silt, not gravel. In the first phase of construction in the spring of 1973, 5000 cubic yards of gravel and sand were transported by truck from a pit at Ya Ya Lake using the frozen river channels as roads to the pit. This material was dumped through the ice and formed the base of the island. In the second phase, begun and completed in October-November, silt from the bottom of the bay was piled on top of the gravel foundation. The silt was held in place by a berm constructed of sand bags. The silt then froze in place and provided the foundation for the drilling rig which was only two feet above sea level. The island was encircled by 90 five-pound explosive charges. They were placed approximately 125 feet out from the island and were suspended 7.5 feet below the ice surface. They were to be fired simultaneously if the monitoring devices indicated that the pressure of ice was causing the island to move. It is not known whether explosives are still routinely placed around islands where drilling is to occur in winter. It is possible that at Adgo it was simply added insurance to protect the island, the first one constructed of silt.

In Adgo's case, the entire drilling operation had to be completed before spring because it was expected that the island would begin to disintegrate with the spring thaw. However, the island did not break up as quickly as anticipated and "was only slightly eroded by storms during the 1974 summer season."[13] Gravel used to construct its base eventually will be reused in the construction of other islands. Adgo was described as an experiment that would allow Imperial to judge the engineering feasibility and economic benefits of this type of construction. The cost of building and drilling from it was about $5 million, about half as much as Immerk. The same design concept has since been used in building Adgo P-25, a subsequent delineation well.

The third island, Pullen, was built by hauling gravel on the ice from the pit at Ya Ya Lake. The gravel was dumped through holes cut in the ice. Thirty trucks were involved in the operation and they hauled an average of 1860 cubic yards a day over a distance of 65 miles. In all, 84,000 cubic yards were trucked from the Ya Ya Lake gravel pits on Richards Island. Construction was completed before the spring breakup, and the rig which was used on Immerk, the Hi-Tower IE, was moved across the ice to Pullen in April. The well, Pullen E-17, was spudded during the spring of 1974 and drilled that summer.

The Unark E-24 island, the first to be constructed by Sunoco, was similar in design to Pullen. Pelly, Sunoco's second island, was somewhat different in design from any of the previous islands. Sunoco, in its application for a land use permit, offered this description: "The island will consist of a steel core, surrounded by silt and protected by bundles of sandbags (gabions) . . . . Hauling and placing of the estimated 3500 gabions required for the island and filling the island with silt is expected to take 6 to 8 weeks."[14]

It appears that Imperial at first anticipated some means other than artificial islands for exploring its permit areas which are under more than 10 feet of water.[15] However by early 1974 it became evident that it was going to attempt to build islands in water much deeper than 10 feet, and applied to build the two Netserk islands, one in 23 feet of water.[16] A story in *Oilweek* announced that "the five biggest dump scows designed and built in North America will be used in island construction for Imperial Oil in the Beaufort Sea this year. The 2000-yard scows are designed to dump their 5000-ton loads by literally splitting longitudinally, activated by hydraulic rams fore and aft."[17] The barges are loaded by six-yard clamshell dredges which are mounted on smaller spud barges.

Again this year there is evidence that Imperial is aiming at island-building at even greater depths, possibly close to the edge of the shorefast-ice zone. In January 1975 Imperial applied for a land use permit to construct Arnak L-30, 13 miles northeast of Pullen Island in 27 feet of water: "The island will contain approximately 1.25 million cubic yards of medium grained sand . . . . The fill will be pumped directly onto the island from a suction dredge similar to the procedure used to construct Immerk B-48 island in 1973-74."[18] The application did not give any detail on the suction dredge, but *Oilweek* reported on it a few months later:

Canada's first ship-mounted suction dredge, Mackenzie Beaver, is crossing the Atlantic under its own steam to catch up to the Western Arctic convoy heading for Prudhoe Bay, on its way to Tuktoyaktuk. The 36-inch suction dredge is mounted on a ship hull, 275 feet long with a beam of 65 feet and a 23 foot draft fully fuelled. Its rated capacity is 50,000 cubic yards per day with a 100 foot discharge. This is close to 50,000 tons.

The Mackenzie Beaver will be primarily island building for Imperial Oil and although it has a suction ladder reaching as deep as 120 feet, an Imperial spokesman doubted whether it will be used for island building in waters significantly deeper than is currently normal with the Northern Construction clamshell fleet. This will depend on ready availability of good sand bars near planned well locations.[19]

But the progression of forecasts on the depths at which islands will be built tends to place this statement in a more realistic perspective, and the rapid buildup of the dredging fleet in 1974 and 1975 indicates that Imperial at least has plans to build many more artificial islands.

For deeper water, Imperial has also considered the construction of retained drilling islands (RDI) and conical drilling platforms (CDP).[20] Imperial has also become a member of the Beaufort Sea Task Force and therefore has a proprietary interest in the studies which were conducted to develop an exploratory drilling system for use in waters up to 200 feet deep.[21] The RDI described at the Northern Canada Offshore Drilling Meeting "would consist of an inner steel barge and an outer steel retaining wall, mutually fastened by means of truss work, leaving an open annulus between the barge and the outer ring. The floating steel structure would be towed to location, ballasted to bottom and the annulus filled with dredged material. The fill material in the annulus and a ring of soil below the sea bed would be artificially frozen by freezing coils and piles. The frozen material in the annulus would provide structural integrity and lateral stability in resisting ice loads. Relocation would involve thawing out the dredged material, deballasting and towing to a new location."[22] The RDI would be approximately 300 feet in diameter.

The CDP concept was described at the Offshore Drilling Meeting as having drilling capacity which might be extended to 120 feet of water. It is a gravity-founded conical structure of reinforced concrete and steel, incorporating a hull and superstructure. The key design feature is the conical shape which would cause the ice to fail in flexure rather than by crushing. The tailing flexure of ice is several fold lower than in the compressive mode: "It would be built on the west coast and towed around Point Barrow complete with drilling equipment. On location, the structure would be set on bottom by controlled water flooding. Sufficient water ballast would be added to attain the necessary stability against ice loads. Relocation, possible only in the open water season, would be achieved by pumping out water ballast, refloating and towing to a new site."[23]

It appears that the current version of the CDP is referred to as a monopod drilling rig or system. At any rate this was the name used to refer to the Imperial system at the Beaufort Sea Investigators' Conference and in a subsequent article on Arctic technological problems:

Structurally, the Monopod is made up of three main components: the hull, the shaft and the superstructure. The total steel work weighs 15,000 tons. The hull is circular for better base pressure distribution and is 320 feet in diameter and 23 feet high. It is protected against ice floes during transit by a peripheral concrete ring five feet thick, which also acts as permanent ballast for increased stability. The shaft is 30 feet in diameter extending 80 feet above the top of the hull to the bottom of the drilling platform. The drilling platform is 160 feet by 80 feet and accommodates in three levels all equipment, machinery and supplies required to drill a 20,000 foot well.[23]

Potential Cumulative Effects of Artificial Islands

The possibility of cumulative environmental effects developing as a result of the continued construction of artificial islands should be thoroughly assessed by DOE's Environmental Assessment Review Programme. This possibility was not considered in the environmental assessment which dealt only with Immerk, the first artificial island.[24] The potential effects on beluga or white whales particularly should be considered. The whales must have open water available to move from area to area, and so preservation of open leads is very important during break-up or in periods when the ice is on the move. It would seem possible that the presence of increasing numbers of artificial islands in a limited area could change the distribution of leads during the critical periods when the whales are migrating to the areas where their young are born. Similarly, the potential cumulative effect of the dredging programme for island construction should be considered. The coastal areas have been shown to contain many spawning and rearing areas for anadromous and marine fishes. Are there any important ones in the areas where the islands are being constructed? if so, could dredging and island building have long-term effects on them? Similarly there should be a thorough public assessment of the potential cumulative effects of the discharge of chemical compounds (drilling mud and associated chemicals) into the near shore areas of the Beaufort Sea. The present policy appears to be that dilution is the solution to pollution.

Blowouts, Contingency Plans, and Environmental Risks

Imperial submitted a specific contingency plan along with its application to drill its Immerk well.[25] Soon after, it adopted the procedure of submitting a standard contingency plan which is kept on file in appropriate government offices and periodically updated.[26] Each copy has a designated number, presumably as a means of controlling distribution of the plan. Sunoco has also submitted a standard contingency plan for its Mackenzie Bay wells. In addition, it included a brief section in its land use application for construction and drilling.[27] Imperial's primary objective in its plan is to prevent blowouts:

Because of the remoteness, inaccessibility and the general logistical problems associated with Arctic drilling operations, it is Imperial's position that problems of extreme magnitude cannot be permitted to occur in the Arctic. However, realizing that a high percentage of oil blowouts results from human factors, the Company has realistically adopted the attitude that the best preventative is to train drilling crew members and supervisory personnel in the latest techniques for handling well kicks.[28]

The contingency plan contains 13 sections, including details on emergency organization, prevention, reporting, preplanning, guide to well control procedures, oil spill control and clean-up, and several sections on administrative procedures, transportation, and supplies. The section on public relations specifies the exact procedures to be followed and the topics to be covered in the event of a blowout.[29] Interestingly, this section of the Immerk plan required a final review of news releases by Imperial's public affairs department and the production department at the head office in Toronto. Normally, the standard plan requires only that the review be done by Calgary officials.

Section 1200 of both the Immerk and the standard contingency plan gives definitions of minor, major, and extreme problems:

Minor Problem: when the inflow of water, oil or gas from the formation is limited and the well can be shut off without fracturing the formation.

Major Problem: when the influx of water, oil or gas takes place and the well can be diverted to flare but cannot be shut in without fracturing the formation; when the length of time that the well can be flared is limited because of the possibility of cutting out the surface control equipment.

Extreme Problem: whenever a significant influx of water, oil or gas occurs and the well cannot be shut in but can be diverted to flare; any influx that cannot be diverted to flare without the resulting back pressure causing the formation to fracture; any condition in which the reservoir fluid is in uncontrolled communication with the surface; when erosional or mechanical failure of the control equipment occurs, resulting in loss of control of the well.[30]

The plan accompanying the Immerk application proposed that in the event of a blowout a second island would be constructed so that a relief well could be drilled. It suggested that probable construction periods would be 1 July to 1 October, and from 15 January to 15 April. The first period might be extended to 15 November if ice-breaking equipment was used. In subsequent applications Imperial has provided more detail on what it would do about the construction of relief islands:

In regard to the question of providing back-up drilling capability in the event of a serious problem with the original well, the following methods would be used where applicable:

(a) From July 15 to September 15, a relief island would be built of dredged materials.

(b) From November 1 to February 1, an ice island would be built, dependent on a location in shallow water as around Adgo.

(c) From February 1 to May 1, a trucked gravel island would be built.[31]

The application for Arnak also lists five items to be considered "so that personnel involved will be completely aware of what action must be taken in the event of an emergency." The items include seasonal wind variation, ice conditions, probable well depths at the time of a blowout, and total depth of the relief well. It states that the relief island would be located within a radius of 2000 feet of the blowout well, however, it might be as close as 500 feet "should the well blow out at a depth of 3950 feet." The submission illustrates "the manner in which the surface location of a relief well would be selected."[32]

In other sections of this book it is pointed out that in the cases of offshore drilling from drillships, semi-submersibles, or ice platforms, the lag-time before a start could be made on the drilling of a relief well ranges from nine months to a full year. It is of some consolation that in the case of artificial islands the lag-time would probably be reduced to a maximum of seven or eight months and a minimum of three or four, depending on the time of year that a blowout occurred. Depending on the depth and location of a problem well, it could still require as much as a year to build the island and drill the relief well.

In the Immerk contingency plan Imperial stated that formation pressures would be normal to 10,000 feet and the well

will not reach a critical stage until after January 16, 1973, which more or less coincides with the start-up of unrestricted movements of heavy equipment over ice surfaces in the general Beaufort area. This well will therefore be accessible for the entire period it is drilling below a depth of 10,000 feet and should a blowout occur below this depth, suitable equipment can be deployed to control the well or alternately, preparations can be made to construct a second island from which to drill a relief well.[33]

But there is evidence that Imperial knew that the well would reach a critical stage at 7600 feet, that is, by late November or early December, six weeks to two months in advance of the period of unrestricted movement of ice in the area. Imperial would have had great difficulty deploying equipment to control the well or to construct a second island if a blowout had occurred; the evidence indicates that the company was aware of that fact.

Major problems can be of much greater significance in offshore operations of any description than a comparable problem at land-based operations. In the latter, oil can be put in the sump or in an emergency containment area. But in an offshore facility, only a limited space and volume is available to contain the liquids which would be discharged while the well is being brought under control.

If a major or extreme problem situation develops, every alternative is considered and tried before the well is set on fire, according to the contingency plan. If such a decision is made, the well is set on fire with a flare gun with range of approximately 500 feet. Before carrying out the procedure, constant checks are maintained to determine if explosive gases are present at the position where the flare gun operator is located.

In the case of summer operations contingency plans are considered on the basis of oil escaping into calm or rough seas. In calm situations containment with a sea-oil boom and the use of skimmers or slicklickers is proposed. If oil escapes into a rough sea the contingency plan admits that "containment and recovery will present a difficult problem to which there is no obvious, easy or practical solution at this time. If a large uncontrolled flow of oil is experienced under these conditions, personnel may have to be evacuated and the well ignited."[34]

The description of procedures for winter operations shows clearly that there is no new technology available to deal with problems which occur at this time of year either. The section on offshore spills during winter states that "The control measures for 'Summer Operations' recommended in Section 1840 also apply in the case of winter operations." There is some possibility that oil which has escaped at offshore locations can be contained in winter but very little possibility of containment at other times of the year. The contingency plan states it this way: "In the offshore case, the emergency containment of produced fluids will only be possible during the winter when snow and ice can be used to construct a containment area on the ice. However, during the summer, or during freeze-up or break-up, containment of such fluids will be impossible and the only alternative will be to ignite and burn such fluids."[35]

This means that an attempt would be made to contain the oil on the ice surface by means of dams and dikes. The section on removal and disposal in the Immerk contingency plan states "burn if possible, scrape or scoop oil-snow-ice mixture with power equipment or by hand into a holding area using hand-brooms and hand shovels for final clean-up." It is difficult to envision effective clean-up with brooms and shovels at -30° to -40°F, a temperature range which is common during December, January, and February. The plan gave this further description of clean-up operations: "Oil spills on the ice must be scraped up as well as possible; oil that cannot be recovered in this way must be collected in depressions on the ice surface and removed with pumps or skimmers depending on the temperatures and the viscosity of the oil. Attempts must be made to recover any oil trapped under the ice by drilling holes with an ice auger."[36]

Reading the sections of the contingency plan which deal with oil spill control and clean-up leaves one with the sinking feeling that the plan is totally inadequate should an oil blowout occur in the Beaufort Sea.[37] The physical problems posed by weather and ice are awe-inspiring, and they are compounded by transportation problems particularly at off-season periods of the year, during break-up and freeze-up. At the Offshore Drilling Meeting an official of Imperial stated, in the last paragraph of a section in his paper dealing with contingency plans, that

Immediate action would be taken to commence the drilling of a relief well. This would mean the building of an island and the mobilization of drilling equipment. The key point to be noted is that once a blowout has occurred, even if a relief rig is immediately available at the site, it is a matter of two or more months before the hole can be drilled to the depth necessary to kill the blowout well. It is therefore really a matter of timing and if containment and burning operations are effective the magnitude of the problem should not become worse with time.[38]

The information on the minimum and maximum period given earlier indicates that Imperial's case was presented in the most positive terms. However, even on these terms, is it reasonable to permit the development of a full-scale exploratory drilling programme in the Beaufort Sea on such a basis? If offshore drilling is to be done in the Arctic should a back-up system, which can be immediately deployed and which can begin to drill a relief well without the long time-lag, not be available for deployment rather than exist only as a futuristic concept portrayed in an engineering journal[39] or in the preliminary phase of an APOA project?[40]

References and Notes

1 The land-use application was submitted by Imperial in November 1971 and approved 22 February 1972. See n. 52, Chapter 8 regarding an apparent misrepresentation of some information submitted with the application.

2 Letter dated 26 June 1973, from C.S. Lewis, Regional Manager, Environmental Protection Service, DOE to Mr. D. Gee, Regional Manager, Water, Lands, Forest and Environment of DINA and a reply on 6 July 1973. This exchange of correspondence and documents associated with the Adgo incident was circulated internally in DOE during 1974 when the department was trying to establish an "effective interface" with DINA.

3 For a more detailed appreciation of the physical environment of the Beaufort Sea, see Chapter 2.

4 Two papers were presented on ice and the sea floor at the December 1972 Northern Canada Offshore Drilling meeting in Ottawa by Dr. E.F. Roots. They portray in dynamic terms these aspects of the physical environment. They were entitled *The sea floor and below* and *Sea ice and icebergs.*

5 Rowsell, K.A., *Notes on some major technical problems,* Northern Canada Offshore Drilling Meeting, December 1972, Ottawa.

6 The distribution of exploration licences in the Arctic is shown on a map which is included periodically in *Oilweek.* The one referred to was published on 20 November 1972.

7 Haight, G.L., *Imperial's offshore Arctic programme,* Northern Canada Offshore Drilling Meeting, December 1972, Ottawa.

8 Based on information obtained from land-use applications and related documents.

9 For more information on formation pressures see Chapter 9.

10 *Imperial Oil discovery greatly enhances potential of crude oil from Arctic regions, Globe and Mail,* 14 March 1974, and *Oil at last makes some good news, Financial Post,* 23 March 1974.

11 *Financial Post,* see n. 10.

12 Articles by Imperial employees on construction of Immerk, and *Oilweek, Imperial Beaufort island building said moving along at satisfactory rate,* 28 August 1972.

13 Imperial, *Beaufort Sea artificial islands,* 16 December 1974, submitted in response to a request for information from the Land-use Advisory Committee.

14 Sunoco EDP Ltd., supporting documentation for an application for land-use permit, 21 May 1974.

15 Haight, see n. 7.

16 Imperial, Application for a Land-Use Permit to construct Netserk North E-30, 16 May 1974.

17 *Oilweek, Mobilization of a $10 million fleet to build islands in the Beaufort Sea,* 1 July 1974.

18 Imperial, Attachment to the Application for a Land-Use Permit Imperial Arnak L-30 — Proposal for Construction, 15 January 1975.

19 *Oilweek, Canada's ship mounted suction dredge heads for island building in Beaufort,* 19 May 1975.

20 Haight, see n. 7.

21 See Chapter 2 for details on the development of a deep-water drilling system.

22 Haight, see n. 7.

23 Brown, A.D., *Arctic offshore drilling — current and future techniques,* Beaufort Sea Investigators' Conference, 21 January 1975, and Jones G.H., *Arctic poses extreme technical challenges, Petroleum Engineer,* January 1975.

24 Slaney, F.F., *Environmental impact assessment, Immerk artificial island construction,* Mackenzie Bay, N.W.T., January 1973.

25 Imperial, *Arctic well control contingency plan,* Immerk B-48, 1 December 1972.

26 For example, Imperial's *Arctic well control contingency plan,* put out by Imperial's Producing Department and periodically updated, 1 January 1973.

27 For example, Land-use Applications for Sun BVX et al Unark O-24 and for Sun BVX et al., Garry P-04.

28 Imperial, see n. 26, Section 1100, "Prevention."

29 See n. 26, section 1250.

30 The definitions in the standard Arctic well control contingency plan were long, technical, and only understandable to a professional oilman; those in the Immerk plan were expressed in more comprehensible terms.

31 See n. 7.

32 See n. 18, Fig. 3.

33 See n. 26 and Chapter 8, n. 52.

34 See n. 26, Section 1840.

35 See n. 26, Section 1340.

36 See n. 25, Section 1430.

37 For a detailed review of oil spill problems see Appendix 6.

38 Haight, G.L., see n. 7.

39 Jones, G.H., see n. 25.

40 Brown, A.D., see n. 25.

Chapter 4

Production and Transportation of Offshore Oil and Gas in the Beaufort Sea

Most of the offshore activity in the Beaufort Sea until now has been exploratory. Drilling to date has been confined primarily to searching for initial discoveries of oil and gas. Once a discovery is made, as in the case of the Adgo oil and gas well, further drilling has been carried out to "delineate" the size of the field. Delineation is an attempt to determine whether or not there is a "threshold quantity" of oil or gas which would make the field economically feasible to develop. The threshold quantity for a particular discovery is based on a number of considerations, including the technology required for production and transport, distance to market, availability of support facilities, and the market price of fossil fuels. Industry has not yet provided definite estimates of threshold volumes for the Beaufort Sea area, but some preliminary information is available.

The threshold volume for Delta gas to be transported economically by pipeline to the south in conjunction with U.S. gas from Prudhoe Bay is estimated by Canadian Arctic Gas Study Limited (CAGSL) to be about 12 tcf. Reserves of approximately 6.5 tcf are classified as proven, probable, and possible.[1] The figure for a threshold quantity of oil in the Beaufort Sea is somewhat more tenuous. Industry suggests that a volume of 2 billion barrels of oil is necessary in order to finance a pipeline from the Mackenzie Delta area, but it is not clear what portion, if any, of the 2 billion barrels would come from offshore. Calculations by the Northern Assessment Group predict that current oil reserves of a little over a billion barrels of recoverable oil would be an optimistic estimate, given 750 million barrels at Adgo (offshore), 300 million barrels at Ivik, 80 million barrels at Niglintgak (onshore), and 200 million barrels at other locations (onshore).[2]

Once a field has been delineated and the threshold level has been proved, a production programme is begun to recover the oil or gas through a number of wells mounted on platforms strategically placed for maximum recovery from the field. Transportation of the oil or gas to shore is normally by undersea pipelines or tankers.

Since oil and gas reserves in the Beaufort Sea are already being considered for development, we will focus on the types of activity that would probably be involved in production and transportation in that area. The methods used and technologies developed for that area will undoubtedly be adapted for use in other regions of the Arctic.[3]

Production Technology — The State of the Art

Oil does not occur in huge pools below the earth's surface. Oil reserves are composed of hydrocarbons found in the interstices of porous sedimentary rock, the most common being sandstone and limestone. The production efficiency from an oil- and gas-bearing geological structure is dependent upon a number of physical variables, such as the porosity and permeability of the geological structures, the presence of pressure exerted by any fluids which are there, the viscosity of the oil, and the depth of the well — all contribute to the flow rate and recovery factor of a reservoir. If the rock contains large fractures or fissures the permeability is greatly increased and a much more efficient level of production can be expected.

This is the case in many Iranian fields and explains the efficiency of these wells compared to fields located elsewhere in the world. Iranian wells are rarely under a mile apart, whereas in other producing fields the wells may be as close as 300 to 400 yards.[4]

Not all the oil or gas that is in a field can be produced for commercial use regardless of the spacing of the wells. It is estimated that the average ultimate recovery for OCS (outer continental shelf) production will be about 45% of the oil estimated to be in place. The figure for Louisiana is 46.5% while that of California is only 23.6%. Overall the average for U.S. production is about 31%.[5] The recovery rate for Norman Wells, the only producing field in the N.W.T., is 10%.[6]

Production wells may be arranged in a regular pattern if the oil or gas-bearing structure is flat, or in a straight line or circle if the formation dips steeply.[7] The number and location of production wells will be influenced by the size of the discovery and whether or not it must be produced from a few large reservoirs or from several small, scattered ones.

Production wells are drilled with the same rigs as exploratory wells, but since they are drilled with the expectation of a continued flow of oil or gas over a long period of time, they are equipped differently. Generally there is no more difficulty involved in drilling production wells than in drilling exploratory wells. In shallow water, where artificial islands have been built for exploratory drilling, production wells will probably be drilled in a similar manner. In the deeper waters of the Beaufort Sea, exploratory and production wells will probably be drilled using the same method. Dome Petroleum plans to drill exploratory wells in depths ranging from 100 to possibly 400 feet of water with two drillships.[8] If they are successful then drillships will undoubtedly be used to drill production wells where the water is too deep for artificial islands. If this is the case, all the deepwater offshore wells will be drilled during the summer. Unless new types of equipment are developed, it will not be possible to conduct blowout control operations during the breakup periods or during the winter. An ice cutter semi-submersible drilling vessel is being developed,[9] but as pointed out in Chapter 5 it will only be possible to use it to drill wells in landfast ice where the water is at least 300 feet deep, a factor limiting its use to the Sverdrup Basin.

After commercial quantities of gas or oil have been found, production wells are drilled and the production facility most suited to the location is installed. The alternatives in most offshore areas are to build either a fixed platform or a subsea unit on the ocean floor. In the North Sea, fixed platforms will be built in water depths of up to 550 feet, and, Exxon is planning to use a fixed platform in 850 feet of water (in the Santa Barbara Channel). Subsea completions, on the other hand, involve placing the wellhead on the ocean floor rather than on a platform. The produced liquids or gases are transferred from the wellhead either to a nearby fixed platform or to a shore facility for processing. In 1973, more than 70 subsea completions were in use in offshore U.S. waters.[10] (In future, such operations could also involve transfers to subsea production systems. These systems are described later in this section.)

Steel platforms anchored to the seabed are most commonly used in the Gulf of Mexico, but in the North Sea concrete production platforms are favoured; they are considered to be more suitable for the stormy North Sea environment. In addition to being more resistant to storms, they are well adapted to the clay and sands of the North Sea ocean floor which are able to support heavy structures.[11] If the same type of unit was used in the Gulf of Mexico, the softer sediments would allow the platform to shift. Production platforms for Cook Inlet in Alaska are made of steel. Most commonly they consist of a deck and tower section which is supported by four large legs through which piles are driven to anchor it to the sea bottom.[12]

Oil and gas have been produced in the Gulf of Mexico for over a quarter of a century and in Cook Inlet for a decade and so some of the kinks have been worked out of the production systems used in these areas. One field in the North Sea has been in production for two years while others are about to start and judgement must be reserved on the reliability of production systems in that area. Oil spills are nonetheless a common occurrence in offshore production systems which have been in operation for a long time.

The Beaufort Sea environment is very different from the Gulf of Mexico, the North Sea, or Cook Inlet in Alaska. Arctic pack-ice would destroy any existing production platform that is in use in any of these areas. Low temperatures would also cause a variety of problems during production, ranging from the working efficiency of men and machines, to the brittle fracture of cold metal, to the growth of ice on metal structures. In the Arctic, ice can build up rapidly from the moisture in humid air as well as from spray. A buildup of ice increases the area exposed to wind, increases heeling movements, and can cause overloading of equipment or of the structure.[13]

Where production wells are drilled in shallow water from islands, these will have to be permanent, unlike those constructed for exploratory wells. They will require more protection from the action of waves and pack ice, probably to be provided by concrete and pilings. Imperial Oil recently made this statement about its production facilities: "All of our islands built to date have been designed only to support exploration wells. It is not intended that they be used as permanent structures. The method of construction for permanent islands is still in the planning stage. These will be designed to maximize the utilization of the exploration islands."[14] As an alternative, or perhaps as an addition, to man-made islands, Imperial has designed a conical monopod-like structure for use in shallow ice-infested waters. The system has received approval in principle from DINA, although no announcement has been made on plans to construct a prototype. The concept was first introduced some years ago in Cook Inlet in Alaska, and since then the Imperial design has been modified to withstand ice loads of 22.5 million pounds of lateral thrust from sea ice (the monopod has been described in more detail in Chapter 3).[15]

In deeper water from 35 to 160 feet, ice scouring will pose a threat to both production and transportation facilities. Here it will not be feasible to use artificial islands or monopods as production platforms beyond that part of the sea where the bottom is scoured by ice; they could not withstand the pressures exerted by ice islands or by the polar pack ice. Subsea production systems will probably have to be used in such deep-water locations. These systems rest on the bottom and operate automatically, and as many as 18 oil or gas wells can be handled by a single system. The state of development of subsea production systems was described in 1973 in these terms:

Two prototype subsea production systems (SPS) are presently being tested under actual field conditions and another is in an earlier design stage, but the subsea production system can be considered a part of existing offshore technology. These systems come in a variety of configurations. They include both wet systems, in which wellhead equipment is exposed to water, and dry systems which contain essentially conventional wellhead equipment within watertight chambers maintained at atmospheric pressure. These chambers are designed to permit workmen to perform maintenance activities in a shirt-sleeve environment.[16]

An SPS reportedly suitable for Arctic operations is being designed by Lockheed Petroleum Services.[17] The subsea production system is maintained by means of an underwater service capsule, which can be lowered onto a wellhead cellar or the manifold center. It is sealed to the unit on which work is being done, and provides a relatively comfortable working environment. The service capsule is connected by a cable and cord which contain the life support system from the surface service vessel. Activities at the production stage has been described as follows:

From the wellhead, the produced fluids are run through various treatment facilities to prepare them for sale or discharge. They pass from the top of the well to the treatment equipment through a group of valves and fittings referred to in the industry as the Christmas tree. The purpose of this hardware is to direct the fluids to treating equipment, to close off the well in the event of malfunction or emergency, to limit the well-flow rate by the use of chokes, and to permit testing and workover or servicing operations to be performed on the well.

The nature of the treatment facilities into which the produced fluids pass from the Christmas tree is determined by the composition of the fluids and the requirements which must be satisfied for the fluids to be sold or discarded. To be discarded, sand must be free of oil, and discharged water must have an average of not greater than 50 parts per million of oil.

If both oil and gas are produced, a separator is needed so that the oil and gas can be metered and pumped through separate lines. If water is produced with oil, a tank to remove water which is not contained in an oil-water emulsion is often used. This tank is known as a free-water knockout. More sophisticated equipment is needed if water is produced as an emulsion with the oil, if sand is produced, or if the gas contains water vapor.

When treatment facilities can be mounted on production platforms, the equipment is similar, if not identical, to onshore components. For subsea production the process may remain eventually unchanged, but the hardware becomes much more complex. It may turn out that the problems of extensive underwater operations are so severe or expensive that it may be preferable to transport the produced materials (gas, oil, water and sand) either to close-in platforms or to onshore facilities for treatment. In either case, the availability of multi-phase pumping systems will be important to the successful expansion of subsea production systems.

Production activities are continuous and require hardware installed for the life of the field. Production problems occur, however, and sometimes the well bore has to be re-entered for various workover or servicing functions. The less serious of these may be handled by using wire-line or pump-down servicing equipment which may be operated from the platform itself. A variety of tools can be introduced into the well bore to carry out functions such as scraping paraffin deposits from the tubing walls, logging data on formations and well behaviour, retrieving and installing downhole safety valves, and fishing for junk left in the hole. More serious problems require workover procedures for which a drilling rig is

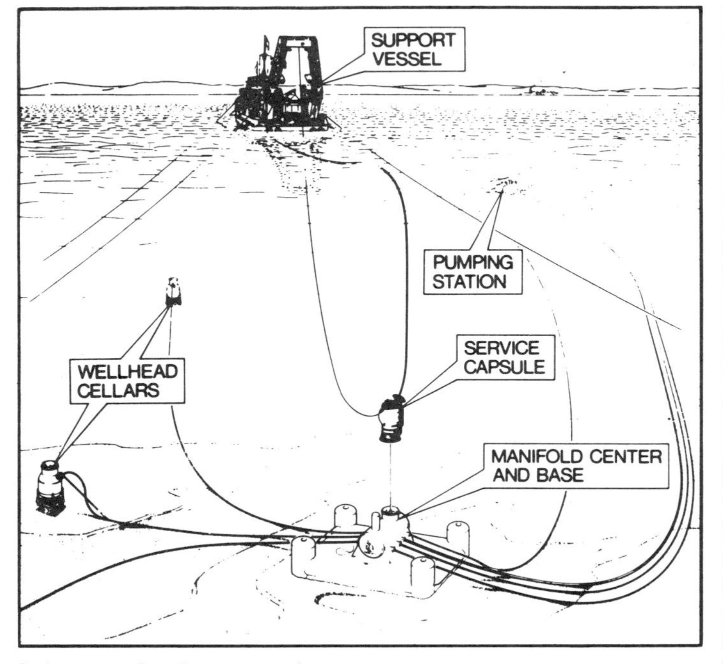

Subsea production system in open water.

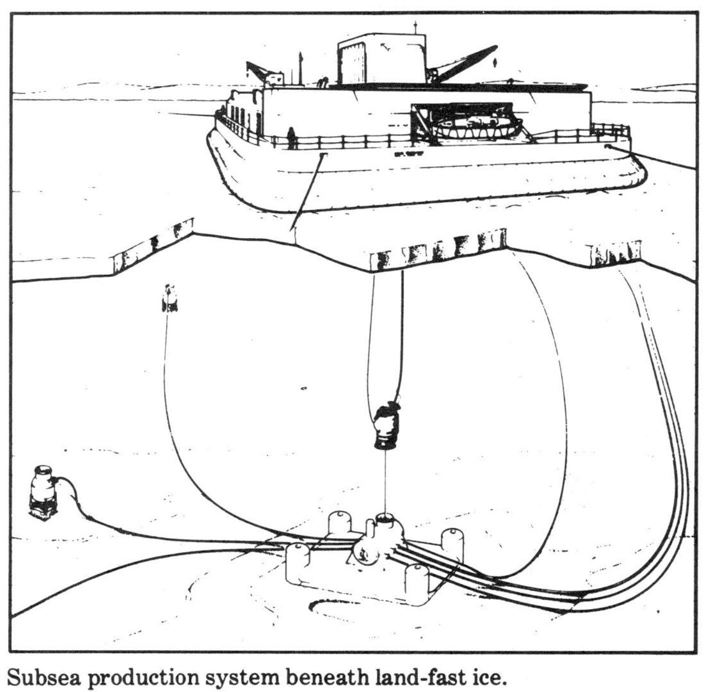

Subsea production system beneath land-fast ice.

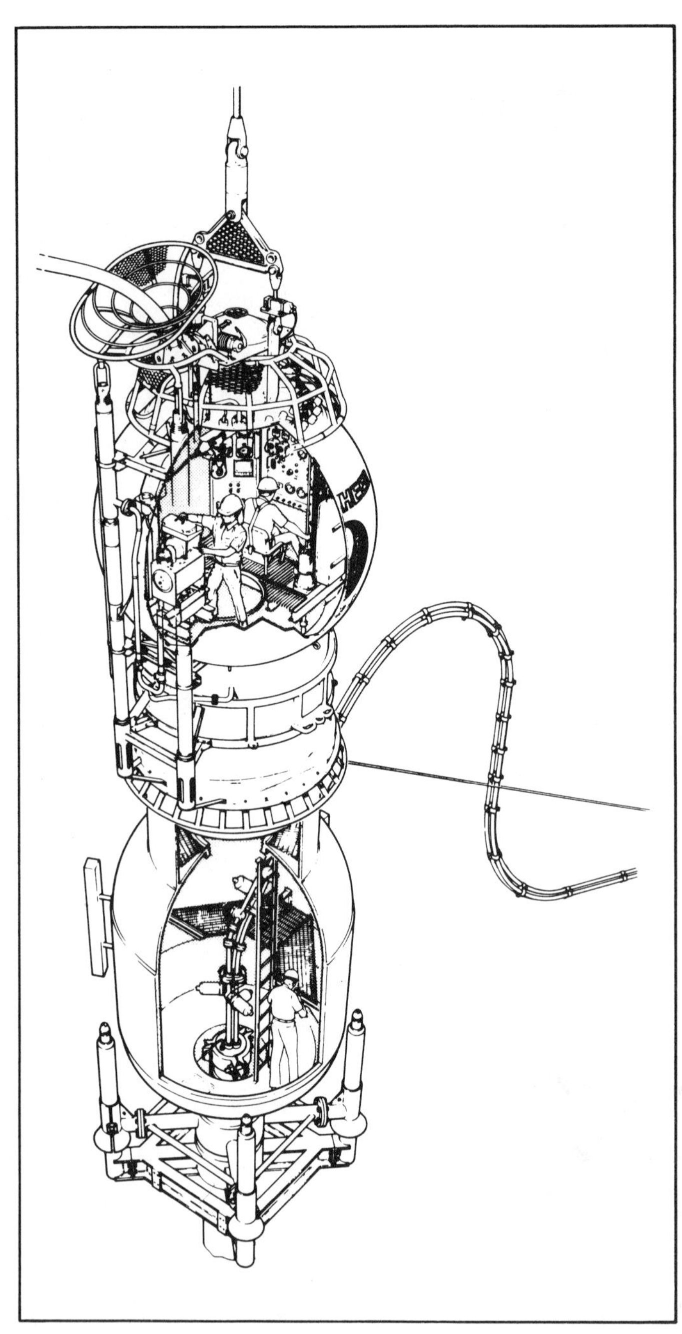

Service capsule sealed to wellhead cellar during work dive.

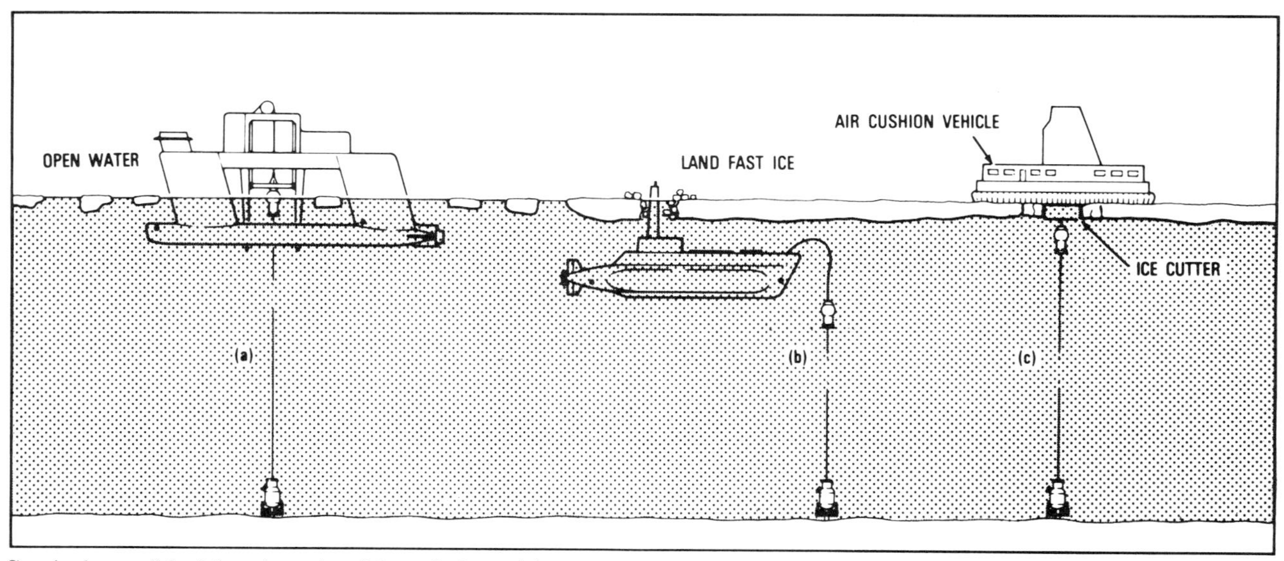

Semisubmersible (a), submarine (b), and air-cushion vehicle (c) are all under consideration as possible Arctic support vessels.

used, and all of the equipment and precautions involved with drilling are employed.[18]

In summary, it appears that artificial islands and monopods will be used for production in shallow waters in the Beaufort Sea. Subsea production systems will have to be used in deep water because it is unlikely that fixed platforms can be designed to withstand the forces which pack ice, ice islands, and pressure ridges could exert against them. It is not very comforting to consider that in 1973 the SPS was still at the prototype stage in the Gulf of Mexico, with the first systems still being tested.

Transportation of Oil or Gas from Offshore

In addition to designing new or improved production equipment for the unique environment of the Beaufort Sea, similar advances in technology must be made in order to get oil or gas to shore. In other areas transportation has been by pipeline and occasionally by tanker. It is unlikely that tankers would be used for transport in the Beaufort Sea, because the Sea is iced-in for at least eight months of the year, because the presence of fog, storms, large ice pans, and the polar pack would be constant threats, and because the system would be very expensive. Consequently, industry will probably depend upon pipelines, as feasibility studies already being carried out suggest.

Pipelines are most commonly laid in the offshore using a "lay barge." As the barge moves along by pulling on anchor lines, sections of pipe are welded together on the vessel and are laid on the bottom with the aid of a support called a stinger. The stinger is required to reduce the stress caused by the weight of the pipe, especially in deep water. A second type of barge, the reel barge, is also used to lay pipe in offshore areas. Using this method, sections of pipe are welded together onshore, wound on a large reel on the barge, and laid out as the barge moves along. This method is commonly used on pipe from four to ten inches in diameter; beyond twelve-inch pipe, the method is not technically feasible. The pipe-pulling method is a third means of laying pipe in the offshore; here, lengths of pipe are welded together onshore and pulled to their destination. This is done by either making them buoyant with floating supports or by giving them negative buoyancy so that they will move easily over the seabed. The limitation of the method is the length of pipe which can be hauled at any one time.[19] Before the pipe is laid, using any method, it is usually treated with a protective coating. Cement is often used on the outside and cathodic devices are also used to prevent corrosion.

A feasibility study for an offshore pipeline in the Beaufort Sea is being conducted by APOA at a cost of $75,000.[20] According to APOA, the purpose is:

To determine the technical feasibility of installing pipelines offshore Mackenzie Delta to the 150' water depth contour. Estimates of installation costs are to be provided in order to establish economic feasibility. Laybarge, Pull, and Reel-barge pipelaying methods are to be considered, limits of technical applicability for each method are to be established and problems identified. Thermal effects of the pipeline will be examined and adequate measures to prevent melting of any existing offshore permafrost will be considered. The study will analyze available scour information, evaluate risk and determine pipeline burial requirements and costs. Trenching techniques form an important aspect of the study. The project is essentially a feasibility study of Arctic offshore pipelines and not a detailed design for a specific line and route. Although certain specific factors in a specific area are being considered, the range of conditions in which a specific recommendation might apply will be given. The sensitivity of any recommended pipeline installation technique to changes of conditions will be indicated.

The physical environment poses three major problems to the construction and operation of pipelines, all of which will require technological adaptations or innovations. The problems are scouring of the sea bottom by ice, the presence of discontinuous permafrost, and pack ice.

In water depths of 35 to 50 feet scouring by ice islands and pressure ridges appears to be intense. In the Beaufort Sea most scours seem to occur in water depths of less than 164 feet. However, there is evidence of past scouring to a depth of 264 feet. Individual scours vary in width from a few feet to tens of yards; the length of some scours is known to be more than 5 miles. About 99% of the scours are 9 feet or less in depth.[21] Since scouring appears to be a cumulative phenomenon, the present course of research is to determine the age and frequency of scouring. Part of the research by members of the Beaufort Sea Environmental Programme (BSEP) has been related to this problem.

As a result of this threat, an attempt will no doubt be made to bury any pipelines below the level of the deepest scour. Jetting machines, which travel on a special carrier on top of the pipe, are most commonly used for burial. As the machine is pulled along, high-pressure water jets are directed at the sea bottom; as the bottom material is loosened, a suction system removes it from under the pipe. In shallow waters dredgers or conventional drag lines are most commonly used. The pipe is then laid directly into the trench which has been excavated.[22]

Although the plan to bury the pipeline seems reasonable, there is still the possibility of a pipeline being crushed by grounding ice. Canadian Arctic Gas Study Limited considered an alternate offshore route in the Beaufort Sea from Prudhoe Bay, Alaska, to the Mackenzie Delta to carry natural gas; they decided that if a 48-inch pipeline encountered grounding ice, it would probably buckle or break.[23] However, feeder lines from the Beaufort Sea will be much smaller and so undoubtedly would be more flexible.

The presence of discontinuous subsea permafrost poses a dual problem in construction and maintenance of pipelines. The presence of frozen materials may require the development of special equipment to excavate the trench for the pipeline. In addition, heat from the oil or gas in the line could melt permafrost ice lenses and cause the pipeline to break or at least be subjected to strong forces. This problem would be common to production systems and pipelines.

Finally, ice on the surface of the sea will frequently make both maintenance and repair operations difficult. Technology is being developed which may eventually handle repair operations in winter under land-fast ice, but repairing a pipeline under the polar pack and along the shear zone of the ice will always be much more difficult. CAGSL considered the problem of pipeline repairs in water 20 to 30 feet deep:

Limited maintenance access presents a serious operational reliability problem. In 20-30 foot water depths, present offshore pipeline repair technology, using surface equipment and divers, is limited to 8-10 weeks of open season. However, the maintenance access in these water depths could probably be extended to 7-8 months per year by the utilization of an air-cushion-supported barge. Such repair vehicles could utilize presently available diver-assisted repair techniques, or a submersible pipeline repair tool presently under development. The latter could result in reduced repair time and eliminate the need of diver support.

However, any type of repair during the break-up and freeze-up would appear not to be feasible, even with the air-cushion-supported vehicle and submersible pipeline repair system. The break-up period extends from the end of May to the end of July and is the most hazardous seasonal condition. The freeze-up period extends from November through January. At present, hazards from moving ice during these two periods prohibit safe operation of most types of support vehicle with the air-cushion vehicle being an exception. However, future ice observations and operating experience during these periods could result in different conclusions.[24]

In the final analysis, CAGSL did not recommend the offshore route in the Beaufort Sea from Prudhoe Bay for a number of reasons, one of which was the significant amount of lead time required to develop adequate technology. Overcoming the technological problems posed by ice-scouring, discontinuous permafrost, and pack ice will not be easy, but with adequate incentive, solutions can probably be found. If large reserves of oil and gas are discovered in Arctic offshore areas, research will undoubtedly be intensified. But even if technical solutions are found, will it be possible to operate production and transportation facilities in the Arctic offshore without the environment being under constant stress from a high level of chronic spills?

Spills of Oil and Gas during Production and Transportation

Spills during production and transportation remain a persistent, chronic problem in the development of offshore petroleum resources. In the Cook Inlet area of Alaska, there are 14 producing platforms, 13 oil and one gas, from which pipelines extend to the shore. In 1968, 26 spills occurred on the platforms. "From 1966 through May 1968, 12 pipeline breaks occurred in the Inlet itself. Some slicks extended many miles and the industry estimates them to have involved more than 1000 barrels of oil. Five pollution incidents have resulted from a single pipeline installation."[25] In total, over 150 pollution incidents were recorded in the Inlet from 1965 to 1968. In November 1967, oil from an unknown source killed an estimated 1800 to 2000 sea ducks and other water birds.

Three major spills have occurred from platforms in the Gulf of Mexico since 1969: "In the Shell accident (1970), estimates of oil lost range from 53,000 to 130,000 barrels. The Chevron accident resulted in loss of 30,500 barrels. Finally the Amoco accident (1971) resulted in loss of 400 to 500 barrels."[26] From 1964 to 1972 there were nine spills from offshore platforms in the United States which resulted in spills of more than 1000 barrels of oil. The range was from 1600 to 77,400 barrels. During the same period there were eight spills of over 1000 barrels from pipelines. They ranged from 1000 barrels of oil to one of 157,000 barrels. In offshore operations in the U.S., which were under control of the federal Government, from 1954 to 1971, one-eighth of a barrel of oil was spilled for every 1000 barrels produced.[27]

The authors of *Energy from the Oceans,* an extensive assessment of offshore technology in the U.S., concluded that each year one in 3000 production wells is involved in a major accident, that is, an accident which results in injury and property or environmental damage. There were 43 such accidents in outer continental shelf operations between 1953 and 1972. They resulted in 56 deaths and in spills which totalled between 290,000 and 1.1 million barrels of oil.[28] The Council on

Environmental Quality (CEQ) summed up the statistical chances of major spills occurring from platforms and pipelines for an oil field of medium size:[29] "There is about a 70 percent chance that at least one platform spill over 1000 barrels will occur during the life of the field. For a small oil field find, there is about a 25 percent chance of one platform spill over 1000 barrels and for a large oil field find, there is over a 95 percent chance of a platform spill over 1000 barrels, during the life of the fields. The probability of pipeline spills follows the general pattern exhibited by platform spill statistics."[30]

In discussing the oil spill problem in Cook Inlet, a U.S. scientist who was familiar with the development of the area, pointed out the fact that industry management "apparently desires to maintain a reputation for good citizenship by running a clean operation. Implementation of this desire at lower echelons is difficult, however." He also cited the problem of prosecuting offenders, because legislation requires that gross negligence be proved. The result was that only five of 150 incidents resulted in prosecutions. He also implied that the safety factors necessary to protect the integrity of oil production and transportation systems against unforeseen conditions were developed by trial and error instead of at the drawing board: "Five pollution incidents stemming from one pipeline in Cook Inlet have emphasized that we cannot afford designs that must be improved upon as a result of repeated failures in the field."[31]

The human factor as a cause of oil spills is emphasized repeatedly in the literature, and it is illustrated in the statistics on the drilling of development wells. According to the CEQ, development drilling is generally less hazardous than exploratory drilling because the characteristics of the geologic formations are better known. However, a survey of 32 wells which had blown out showed that 65% of the wells had been development wells; because drillers act with more caution on exploratory wells, there is an extra margin of safety which makes the difference.[32]

References and Notes

1 The value of 6.5 tcf of natural gas as proven, probable, and possible reserves is the figure calculated by J.C. Sproule and Associates Ltd., December 1974, using the following definitions:

1) "Proven reserves" are considered to be those reserves which, to a high degree of certainty, are recoverable at commercial rates.
2) "Probable reserves" are considered to be those reserves which may be reasonably assumed to exist because of geophysical or geological indications and drilling done in regions which contain proven reserves.
3) "Possible reserves" are considered to be less well defined by structural control than probable reserves and may be based largely on electrical log interpretations and widespread evidence of crude oil or gas saturation. They also may include extensions of proven or probable reserve areas where so indicated by geophysical or studies.

In *Estimates of the Natural Gas Reserves and Deliverability, Mackenzie Delta Area, Northwest Territories* (as of December 1974), prepared for Canadian Arctic Gas Pipeline Ltd., December 1974 by J.C. Sproule and Associates Ltd., Oil and Gas Engineering and Geological Consultants, Calgary, Alberta.

2 Shearer, J., personal communication.

3 Several papers have appeared recently on technological aspects of petroleum development in the Arctic, e.g., McGhee, E. *Drillers weigh their options for the ice-covered Arctic seas, Oil and Gas Journal,* 6 May 1974, also Jones, G.H., *Arctic poses extreme technical challenges, Petroleum Engineer,* January 1975, and Mason, B.N., *Arctic subsea completions, Petroleum Engineer,* January 1975.

4 British Petroleum Co. Ltd., *Our industry, petroleum* (London: Britannica House, 1970).

5 Kash, D.E. et al., *Energy under the oceans,* (Norman, Oklahoma: University of Oklahoma Press, 1974), Ch. 3.

6 Oil and Minerals Division, DINA, Ottawa, 1975.

7 British Petroleum Co. Ltd. See n. 4.

8 For details see Chapter 2.

9 *Arctic tests of ice cutter concept proved feasibility of year round mobility, Oilweek,* 19 May 1975.

10 Kash, see n. 5.

11 U.S. Department of Committee on Commerce, *Outer continental shelf oil and gas development and the coastal zone, National Ocean Policy Study* (Washington, D.C.: U.S. Government Printing Office, November 1974).

12 Visser, R.C., *Offshore oil and gas field development, Cook Inlet, Alaska,* Northern Canada Offshore Drilling Meeting, December 1972, DINA, Ottawa.

13 Dahl, P., *Ice growth will hamper Arctic offshore platforms, Oil and Gas Journal,* 21 April 1975.

14 Imperial Oil Ltd., *Beaufort Sea Islands,* 16 December 1974. Submitted in response to a request for information from the Land Use Advisory Committee.

15 See also Brown, A.D., *Arctic offshore drilling — current and future techniques,* paper presented to Beaufort Sea Investigators' Conference, January 1975 and Jones, G.H., see n. 3.

16 Kash, see n. 5.

17 Mason, B.N., *Arctic subsea completions, Petroleum Engineer,* January 1975.

18 Kash, see n. 5.

19 British Petroleum Co. Ltd., see n. 4.

20 APOA Project No. 39.

21 Pelletier, B.R., and Shearer, J.M., 1972. *Sea bottom scouring in the Beaufort Sea of the Arctic Ocean,* 24th International Geological Congress, Section 8, 1972; also Shearer, J.M., and Blasco, S., *Further observation of the scouring phenomena in the Beaufort Sea,* Geological Survey of Canada Paper 75-1A, and Shearer, J.M., personal communication.

22 British Petroleum Co. Ltd., see n. 4.

23 Canadian Arctic Gas Pipeline Ltd. *Alternate corridors and systems of transportation,* Section 14 e, Subsection 1.3, An Exhibition in Support of Applications to the Department of Indian Affairs and Northern Development of the Government of Canada for Authorization to Use Land to the National Energy Board of Canada for a Certificate of Public Convenience and Necessity for Authorizing the Construction of Pipeline facilities.

24 Canadian Arctic Gas Pipeline Ltd., see n. 23.

25 Evans, C.D., *Environmental effects of petroleum development in the Cook Inlet Area,* in *Proceedings of the 20th Alaska Science Conference,* 1970.

26 U.S. Council of Environmental Quality, *OCS oil and gas — An environmental assessment.* Superintendent of Documents (Washington, D.C.: U.S. Government Printing Bureau, 1974).

27 Kash, see n. 5. One line of speculation is that the Adgo field in the Beaufort Sea contains between 750 million and one billion barrels of oil. If the spill rate of 1/8 bbl/1000 bbls produced prevailed, then over the life of the field between 93,750 and 125,000 barrels of oil would be spilled as the result of chronic accidents.

28 Kash, see n. 5.

29 A medium find was defined by MIT as 2 billion barrels of oil in place and a gas/oil ratio of 1000:1. A small find was defined as 500 million barrels of oil and 500 billion cubic feet of gas and a large find as 10 billion barrels of oil; Kash, see n. 5.

30 Council on Environmental Quality, see n. 26.

31 Evans, C.D., see n. 25.

32 Kennedy, J.L., *Losing control while drilling: A 32-well look at causes and results, Oil and Gas Journal,* 20 September 1971.

Chapter 5

Drilling from Ice Islands in the Arctic Archipelago

Major offshore drilling operations in the next decade will not be confined to the Beaufort Sea region. Using quite different technology and facing vastly different environmental conditions, oil companies engaged in the companion High Arctic "exploration play" also plan to conduct extensive and widespread offshore operations in the Arctic Archipelago.

Led by Panarctic — the 45% government-owned consortium which operates exclusively in the High Arctic — operators in this region have already discovered far greater reserves of gas than have thus far been found in the Mackenzie Delta and Beaufort Sea. Firms operating in the remote regions of the Arctic Islands generally have escaped even the limited public concern focused on operators in the Delta region. And in spite of the federal Government's predominant interest in Panarctic, the consortium's attitude to public disclosure of information related to the environmental aspects of its operation is no different from any of its older and more established counterparts in the oil industry.

According to a Panarctic official's estimate, 75% of the oil and gas potential of the High Arctic lies offshore.[1] But the offshore drilling memo presented to Cabinet from DINA in July 1973 made only passing reference to offshore drilling in the Sverdrup Basin and other areas of the Arctic Archipelago, although the Government officials who sit on Panarctic's Board of Directors must have known of the company's plans to push into offshore regions. Nor was the subject of drilling in these areas considered at the Northern Canada Offshore Drilling Meeting held in Ottawa in December 1972.

Up to that point the only public hint of plans for offshore drilling in the High Arctic was given by Charles Hetherington, president of Panarctic, in an interview for *Oilweek*. At one point in the interview he was quoted as saying that offshore development drilling would have to wait until equipment to drill offshore wells became available, but later said that offshore drilling might become possible with conventional equipment. The scenario he developed was the one followed to drill Hecla N-52 a little more than a year later.[2]

The first public information of any substance on Panarctic's offshore ventures was not released until 1 April 1974 with the announcement that the consortium had been successful in finding gas at its first offshore well, officially named Panarctic Tenneco et al. Hecla N-52 ("Offshore Hecla").[3] Drilled from a floating "ice island" 425 feet in diameter and 17 feet thick at the centre, Offshore Hecla was a delineation well, designed to confirm the extension of the Hecla gas field out under the ocean floor. The well was located about eight miles from shore in Hecla and Griper Bay to the west of the Sabine Peninsula of Melville Island.[4] The Hecla field, which may well be the largest reserve in the High Arctic to date with more than 1 tcf of gas, was discovered in 1972 and was Panarctic's fourth major gas strike. Prior to Offshore Hecla, one successful delineation well (Panarctic Tenneco I-69) had been drilled from a sand spit three miles north of the discovery well (Hecla F-62). To pin down the exact dimensions of the field, further offshore delineation wells may be drilled as far as 30 to 40 miles out to sea.

The Environment of the Queen Elizabeth Islands[5]

The Queen Elizabeth Islands lie north of the great east-west passage which transects the archipelago from M'Clure Strait to Lancaster Sound. They are Canada's isles of the polar sea, the most northerly lands in North America.

As a group, the Queen Elizabeth Islands are generally triangular in arrangement, with the base fixed along the passage mentioned above. To the north-west, the islands extend to the Beaufort Sea, from Prince Patrick Island to north-western Ellesmere Island. To the east, the islands are fringed by Baffin Bay and several passages which separate them from north-west Greenland. The principal islands in the group with their areas in square miles are as follows: Ellesmere 82,119, Devon 20,861, Melville 16,141, Axel Heiberg 15,779, Prince Patrick 6,081, Bathurst 6,041, Ellef Ringnes 5,139, Cornwallis 2,670, Amund Ringnes 2,515. In all, the land area of the Queen Elizabeth Islands totals 165,500 square miles.[6]

The fact that the islands are separated and surrounded by so much water is a determining factor in the general climate of the archipelago area. In the winter months the inter-island passages are frozen over to create, in effect, a single unit mass of snow and ice continuous with the islands. As a result, a continental climate is created during more than seven months of the year.[7] Mean January and April temperatures of -34°F and -18°F at Eureka, -31°F and -14°F at Isachsen, and -27°F and -9°F at Resolute indicate the nature of the winter climate (Table 1, Chapter 8):

Only one Arctic station in two has a record low temperature colder than -60°F, and several have never reported temperatures as low as -50°F. These temperature extremes reflect the moderating influences of relatively warm water beneath the ice-covered channels.[8] *Concerning winds in the area, a surprising feature of the wind pattern over the Archipelago during this period (December to April) is the large percentage of calms reported at most stations. Calm conditions occur almost 30 percent of the time at Isachsen, Mould Bay, Eureka, Resolute and Frobisher Bay and 45 percent of the time at Alert. At these stations winds are light (under 10 miles an hour) more than half the time. Hourly wind speed has exceeded 60 to 70 miles an hour at most locations and several stations along the exposed eastern flank of Baffin Island have reported winds of 100 miles an hour.*[9]

The climate is further described as follows:

While eastern sections of the Arctic, in particular, may be subjected to substantial variations in temperature from year to year, large rapid temperature fluctuations during a particular month or season are uncommon. During the season of continuous ice-cover in the seas and channels, the Arctic is relatively cloud free. Although low pressure areas occasionally cross the region, the cold air is too dry to permit formation of effective snow-producing clouds and, as a consequence, snowfall is very light. Although the steady Arctic cold and light snowfall are characteristic features of the winter climate, it is only when they occur in combination with strong winds that travel becomes hazardous or, in the case of heavy blowing snow, even impossible. The most uncomfortable area and the region where blizzards are most frequent is not the High Arctic, but the coastal sections of the eastern Arctic and the barren lands surrounding Hudson Bay, where cyclonic activity is greater and strong winds more frequent than elsewhere in the Arctic.[10]

The summer months are cool with uniform temperatures of less than 45°F at most coastal locations and only rarely exceeding 65°F for brief interludes.[11] In general, rainfall is usually limited to one or two inches per year in the Queen Elizabeth Islands except for perhaps the east coast where high mountains account for a slightly increased rainfall: "During July and August the maritime influence of the seas and channels surrounding the Arctic Islands stands out as a major control of the climate. At Resolute, Mould Bay and Isachsen and at most other coastal stations, cloud ceilings are below 1000 feet and/or visibilities below three miles about 30 percent of the time during this period."[12] As a result, "in June, July and August, low-lying stratus clouds and coastal fogs are notorious features of the climate."[13] In September, shorter days bring the return of winter: "Over the Queen Elizabeth Islands, mean daily temperatures are below 32°F by the beginning of September and by the end of the month temperatures throughout the Arctic are below freezing."[14]

With long cold winters and cool brief summers, it is not surprising that ice conditions for the High Arctic islands are generally heavy:

The channels of the Queen Elizabeth Islands present ice conditions for surface navigation second in difficulty only to those of the Arctic Ocean. Throughout most of the straits there is continuous ice cover at practically all times of the year, and only in the more favourable summers are there openings in the form of small leads. Ice conditions are best in the east. Parry Channel, which forms the southern margin of the area, has relatively clear summer conditions in Lancaster Sound, but these deteriorate further west [15] *. . . .*

Ice observations in the western channels of M'Clure Strait, Ballantyne Strait and the waters surrounding the Ringnes Islands have shown that there is virtually no movement of ice during the winter and very irregular drift in summer[16] *. . . . Due to the effects of storms the sea ice in the larger channels occasionally loosens up in winter, but in most passages freeze-up begins in October and breakup occurs in June or July. Local bay ice may be later in freezing itself. Sea ice may vary from six to eight feet in thickness, but some floes from the polar pack, according to reports, may be thirty feet thick, and segments of shelf ice from north-west Ellesmere Island, which are not of uncommon occurrence, may be considerably thicker. Well supplied with enormous bergs from glacial discharges along the Greenland coast and supplemented by local calving along the coast of east Sverdrup Land icebergs are common along the east coasts of the Queen Elizabeth Islands.*[17]

Ice island movements in the channels and the distribution of driftwood suggest a

general southeasterly and northerly movement of surface water from the Arctic Basin through the Archipelago. Stronger local currents result from tides although the low tidal range must limit them except in narrow channels. The highest tides are found in the southeastern part of the area where at Resolute the maximum tidal range is about 6 ft. The range decreases westward through Parry Channel until along south Melville Island, it is about 4 ft and on Prince Patrick Island 1.2 ft. Similar low ranges occur in the more northerly islands of the Sverdrup Basin.[18]

Below the water level, knowledge of the environment is more limited. In the words of Canada's Dominion Hydrographer, "While it is true that the earlier problem of geographically misplaced islands and land masses has been mainly overcome, the bathymetric information over much of the Arctic is acutely scarce."[19] The channels between the islands "are interpreted as drowned river valleys. They have been much modified by glacial action and they characteristically have steep trough sides and a horizontal, although somewhat irregular, floor is separated by rills into basins. The general depth in the channels is about 1300 ft."[20]

Hydrographic information shows that "the greatest depths within the Queen Elizabeth Islands occur on the troughs that have been discovered in Parry Channel and the Prince Gustaf Adolf Sea. The depressions, which in all cases are aligned parallel to the long axis of the channels, may reach a depth of over 700 meters and are believed to be the result of glacial deepening of the valley floor at a period when the land was at a considerably higher elevation than at present."[21]

One of the most productive regions of the seas of the world, perhaps the most productive of all, is the Antarctic Ocean, and certainly the least productive is the Arctic Ocean, yet the Arctic and Antarctic have similar temperature regimes. To account for this difference, Arctic oceanographer Max Dunbar explains that "in the final analysis the determining factor in productivity is the degree of vertical instability of the water column. The greater the vertical stability, the less the biological production . . . and in the Arctic Ocean, which is a contained ocean and has limited means of exchange with the other oceans of the world, the freezing and melting of the surface layer cause progressive dilution which also makes for high density stratification and therefore high stability, hence chronic low productivity."[22] As a result, "it is now almost certain that any marine region which contains only Arctic water, that is to say water formed in and coming from the upper 200 meters of the Arctic Ocean, will be very low in biological production and therefore not commercially exploitable for food."[23]

Biologically, the marine environment of the Queen Elizabeth Islands can be divided into roughly three geographic areas, the eastern Arctic, the western Arctic, and the Parry Channel area consisting of M'Clure Strait, Viscount Melville Sound, Barrow Strait, and Lancaster Sound. Of these, most aquatic animals appear to frequent the eastern rather than the western Arctic.

Seals in the eastern sector of the Queen Elizabeth Islands are an important food source for the inhabitants of that area. Ringed and bearded seals have been reported as common as far north as Alert at the northern tip of Ellesmere Island.[24] The harp seal also uses the entire east coast of Ellemere Island as a year-round habitat. The number of breeding grounds, staging areas, and travelways diminish, however, from the east to the west boundary of the island complex.[25]

Polar bears appear to maintain specific areas of concentration. Among the areas of the Queen Elizabeth Islands, the most important in this respect are the north end of Coburg Island (mouth of Jones Sound), shore of Devon Island, a large section in the southern portion of Norwegian Bay, and a small part of western Bathurst Island. Walrus and whales appear to be more plentiful in the eastern sector of the Queen Elizabeth Islands. The walrus, beluga, and narwhal are principal summer inhabitants of the eastern waters.[26]

Waterfowl use several sites of the northern islands, particularly in the eastern part, for summer breeding or staging areas, one of the most important areas being on either side of Eureka Sound.[27] Even the most northerly tip of Ellesmere Island is a critical waterfowl breeding area.[28]

Panarctic's Offshore Drilling Operations

The technology used in drilling from an ice island was, according to Panarctic, a logical extension of the technical approach used the preceding winter to drill four "stratigraphic holes" directly from the ice in Kristoffer Bay off Ellef Ringnes Island. These wells were shallow probes (from 100 to 1600 feet) drilled on 6 to 8 feet of ice and in water depths up to 290 feet. In addition to verifying the drilling procedures their purpose was to check the thickness of permafrost layers in order to assist in the interpretation of seismic data covering the area.[29]

Although Panarctic has maintained that these stratigraphic tests proved the viability of drilling directly from the ice, the Kristoffer Bay tests were not without problems. At one of the Kristoffer Bay sites, the drilling crew experienced "loss of hole" which caused them to "skid" the entire rig.[30] This problem was apparently caused by operations on the ocean floor associated with the installation of blowout prevention equipment during the drilling of the first section of the test hole.

As a result of this problem and their belief that no oil or gas would be found in the upper sediments of Hecla and Griper Bay, Panarctic engineers requested that they be allowed not to use this equipment during drilling of the surface hole at Offshore Hecla. In fact, Panarctic was even prepared to drill a preliminary stratigraphic hole to prove to the Government that no oil or gas would be encountered in the first 500 feet and thus that blowout prevention equipment would not be required during this phase of drilling the delineation well.

The Offshore Hecla delineation well presented engineering problems of a much greater magnitude than the stratigraphic tests on Kristoffer Bay. Whereas the shallow stratigraphic holes had been drilled by a light 150-ton rig, Offshore Hecla required a much heavier rig of 500 tons to support the rig and ancillary equipment. The drilling platform was, in effect, an ice-locked floe. The platform was built up to a maximum depth of 17.5 feet by successive floodings. The thickness of the resulting "ice island" reached its maximum at the "moonpool" (where the drill was lowered into the ocean) and tapered off until, at the edge of the 400-foot diameter floe, it blended uniformly with the surrounding 7.5-foot thick sea ice.[31]

Much depended on the outcome of the Offshore Hecla well. For Panarctic, successful completion of the well would vindicate several years of research that had begun with ice strength tests and the shallow stratigraphic holes. The assumption was that successful completion would suddenly shift the ice-island technique from the frontiers of technology into the realm of accepted and conventional engineering practice. This assumption was clearly set out in the company's 1973 annual report: "Panarctic has pioneered many technological advances in Arctic operations over the years but probably the most significant came in 1973. This was the adaptation of conventional drilling platforms . . . the importance of this method is that offshore drilling can be done in the Arctic now for about one-quarter the cost of employing other proposed techniques which cannot be avaiable for another four or five years."[32]

According to the same annual report, the next step in Panarctic's offshore game-plan was "to use the same 500-ton rig for drilling 6000-foot wildcats in the spring of 1975." Panarctic, in fact, applied to use the same ice-island technique at a well designated as Jackson Bay G-16 during the winter of 1974-75. Government officials, certainly those three senior officials who sit on the Panarctic Board, were well aware that the Hecla offshore test was considered a precedent for an entire subsequent programme of offshore drilling in the High Arctic. Consequently, the Hecla operation ought to have been subjected to an exhaustive review at least within government circles, if not in the public sphere.

Preparation for the Drilling Programme

Panarctic made an application to DINA for approval in principle of its drilling programme on 10 July 1973. It was supported by a report which "was prepared to provide technical support to an application for a drilling authority to drill expend-

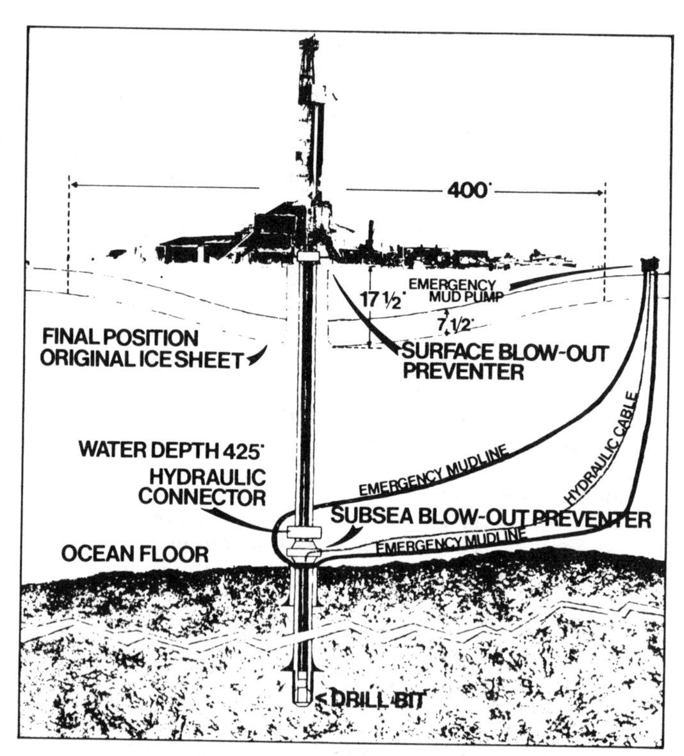

Panarctic Oils built an "ice island" to support its Offshore Hecla drilling operation. The ice platform was built up by flooding the sea ice until a 17½-foot thick "island" had been created, capable of supporting the 500-ton drilling rig. The "ice islands" by nature are temporary and drilling can only take place during the coldest part of the year, from February to April. Courtesy of Oilweek.

able delineation wells from the ice offshore in Kristoffer Bay and off both the west and east coasts of Sabine Peninsula in the Arctic Islands, N.W.T."[33] A study on ice movement[34] included data for the spring of 1971, indicating that planning for ice island construction must have been underway in 1970. The report included brief and very general sections on geology and environmental conditions of the area and a one-page contingency plan for a gas blowout. The remaining three sections dealt with the drilling support system, drilling equipment, and drilling operations. The appendices included data and the theoretical formulations on which the report was based. No information was provided on the ecology of the area, and no reference was made to an assessment of the environmental impact of the operation.

A meeting about the project was held between Panarctic officials and DINA's Oil and Minerals Division, on 28 August. Panarctic was requested to provide additional information on three engineering topics, on an oil spill contingency plan, and an environmental impact statement.[35] Panarctic responded quickly and submitted a supplementary statement on 10 September including a report entitled "Initial assessment of environmental effects and proposed monitoring programme."[36] The letter of submission from the consultant, F. Slaney and Co. Ltd., stated that "It represents the best judgment of our northern staff experts that could be developed with available time and data."[37] The letter was dated 7 September 1973.

The topic came before the Management Committee of the Department of Environment for preliminary discussion on 6 September. The Committee agreed that "the operation as proposed by Panarctic is novel, the unknown factors are even larger than in the Beaufort Sea, and the environmental concerns are therefore significant. The Committee . . . agreed that a formal request to DINA should be made for an environmental assessment of the Panarctic proposal."[38]

Environmental Hazards and Risks

In its submission for approval in principle, Panarctic confidently maintained that drilling Offshore Hecla would not constitute a threat to the environment. On the basis of its seismic and stratigraphic data, which it claimed had proven to be exceptionally accurate, the company estimated that it would hit gas sand at 3000 feet.[39] It was on the strength of this belief that the original submission failed to include a contingency plan for an oil blowout. Panarctic also noted that formation pressures (which had forced Imperial Oil to discontinue drilling on the artificial island Immerk) were well known at Hecla.

The environmental impact assessment outlined the sequence of events that would occur as a result of a gas blowout. It predicted that gas would vent under high pressure and would result in the rapid destruction of the ice platform. The drilling rig would collapse and sink to the ocean bottom. It was predicted that about 50 million cubic feet of gas per day would escape to the surface.

The direct environmental effects of such a release of gas to the surrounding waters would depend on the nature of the impurities in the gas and particularly on the percentage of hydrogen sulphide and hydrocarbons with high molecular weight. Even in minute concentrations, these are highly toxic to aquatic life. Panarctic's data, however, indicated that gas in the Hecla field was very low in these impurities. In such an accident, a further danger would be posed by the loss of fuel, drilling mud, and other chemicals at the well-site. Panarctic's initial contingency plan did not include any measures to deal with this problem although this was later altered at DOE's request. The potential effects of the discharge of large volumes of methane gas into the atmosphere, burned or unburned, is unknown. Presumably, Panarctic must have considerable data on these matters as a result of the gas blowouts at its King Christian and Drake Point wells.

An oil blowout from any offshore well drilled directly from the ice surface poses far more serious environmental hazards than a gas blowout. During all the negotiations prior to the drilling of Hecla, Panarctic took the position that there was only the most remote chance that the well would encounter oil. Such reassurances will be quite impossible, however, when Panarctic uses the ice-platform to drill wildcat wells. Compared to a delineation well, a wildcat well poses several unknowns to the driller: there is far less certainty, if any, of the depth of potential hydrocarbon-bearing strata; there is less indication, if any, of whether discoveries will be in the form of oil or gas; and there is little evidence, if any, to indicate the presence of abnormal geostatic pressures. Panarctic acknowledged that abnormal geostatic pressures had been encountered on three different islands prior to 1972, so risks to wildcat operations do exist from that source.[40]

Panarctic's Hecla contingency plan required that oil from a blowout be ignited as soon as possible. The plan predicted that this could be done readily because of the natural gas which would be associated with the oil. However, there is far from universal agreement that igniting the oil would be possible. Should oil vent from a number of fractures rather than from the drill hole in the formation, it might well be impossible to set it afire; secondly, it has been suggested that it might be difficult to keep the oil burning at the extremely low temperatures common in the Arctic winter, even though there might be some gas present. Even successful burning might pose problems. It could, for example, result in premature break-up of ice making it much more difficult to prevent oil from spreading throughout entire bodies of water during the summer. But since there has been no experience in cleaning up oil discharged from offshore wells in the Arctic, contingency plans and criticisms of them are of a highly speculative nature.

The environmental hazards associated with an oil blowout offshore would, of course, be entirely different than those of an offshore gas blowout. The environmental impact assessment offered only a very general account — and one that many scientists would hotly dispute — of the potential effect on marine life:

Lighter components will evaporate if exposed to air but at a lower rate in the Arctic because of cold temperatures. Biodegradation due to oxidation by bacteria, yeasts and fungi would be limited and slow. Water in oil emulsions could form readily because of low temperatures.

Little can be documented about effects on fish associated with oil pollution in the High Arctic marine environment. No oil has yet been discovered near the site, thus the possible chemical composition of any oil is in the realm of speculation. The species composition, distribution, relative abundance, and spawning and rearing areas of fish inhabiting the Bay are unknown. There is no doubt that a major oil blowout would affect fish populations within the Bay through direct physical effects, and direct and indirect toxicity. The area of impact would be relatively small, however, given effective containment and cleanup. An oil blowout would have serious effects on both pelagic and benthic invertebrates near the blowout site. Many aromatic hydrocarbons are toxic, and would in all probability have lethal and sub-lethal effects on many organisms. Also, there would be direct physical effects on those pelagic organisms which come into contact with spill by inhibiting locomotion and/or preventing feeding. Similarly, emulsification would occur and heavier fractions would settle to the bottom where benthic organisms would be exposed to the same toxic and physical conditions. Sea birds such as the eiders and old squaw duck would be affected by oil in any open water during July, August and September. However, bird involvement should be minimal because spill is unlikely to approach shallow waters or shorelines. Numbers are unknown. A colony of brant near the southern end of Hecla and Griper Bay would not be affected. Chances of a slick penetrating that distance through the sea ice are remote.

Unfortunately, Slaney's entire assessment was stated in the same general terms. It seems obvious that it was prepared entirely on the basis of library research. Considering that the assessment was prepared for an operation which clearly had as a major objective to test ice-drilling projects' feasibility, it is a distressingly inadequate document. Slaney's letter to Panarctic accompanying the assessment maintained that

Even without a full complement of environmental data specific to the region the report represents adequate assessment for the following points:

1. *the project would cause direct environmental effects of only minor nature if operations are normal and environmental safeguards fulfilled;*
2. *a monitoring programme would provide data for understanding implications of ice drilling on a larger scale;*
3. *effects of a natural gas blowout would be minor;*
4. *an oil blowout would not have extensive environmental consequence when contained by burning; the consequence of an oil blowout would not interfere with current resource exploitation activities of native people.*

It is difficult to understand the justification for Slaney's conclusions about the impact of oil and gas blowouts. What justification is there for predicting a "minor" impact when baseline data on the area are almost completely lacking? And the point that an oil spill would be contained by burning is both crucial and doubtful. DOE scientists certainly took a less sanguine view of an oil blowout's consequences. An internal departmental report stated flatly that an oil blowout at Hecla would be "an environmental disaster of the highest magnitude."[41]

The 1974-75 Drilling Season

Panarctic applied for approval to drill two more wells from ice islands during the 1974-75 drilling season. The proposals were for a delineation well, East Drake I-55, for the Drake Point gas field on the Sabine Peninsula of Melville Island,[42] and for a wildcat well "in the Jackson Bay area north of King Christian Island on the west side of Ellef Ringnes."[43] The site is approximately five miles offshore and in 300 feet of water.

The East Drake well was drilled in 425 feet of water from an ice platform of the same dimensions (17 feet thick, 425 feet in diameter) as Offshore Hecla. A second ice platform was constructed to accommodate a larger rig in case a blowout occurred. The well was successful and established that the Drake Point gasfield was more than 25 miles long.[44]

Panarctic submitted an application to DINA in June but was not given approval in principle for the Jackson Bay wildcat well until 4 October 1974.[45] The approval was subject to specific conditions which were outlined in a separate document.[46] However, the second paragraph of the letter, while stating a condition of the approval, made a noteworthy admission on the state of knowledge on the environment:

The approval to drill a well is conditional on Panarctic Oils Limited's ability and willingness to furnish the necessary environmental data to enable the impact of the drilling operations on the Arctic environment to be fully assessed. Current information on the effect of drilling operations on the Arctic environment is insufficient and the proposed operation is considered to be an opportunity, for both industry and government, to expand the state of knowledge of the impact of such activities on the Arctic environment.

The "attachment" required Panarctic to "conduct studies designed to provide data" in the following areas: "Baseline environmental and resource data on fish, marine mammals and birds in or utilizing the Jackson Bay area and in any areas that might be affected by the drift of an oil slick; detailed ocean current studies in Jackson Bay; monitoring for environmental changes during the after drilling operations."[47]

There are several confusing elements in the Jackson Bay case history which are difficult to clarify. For example, it is difficult to understand how the terms of reference for the environmental study were related to the study which was conducted. Sometime during the spring of 1974, at least five months before the approval in principle was granted and environmental conditions stipulated, Panarctic had given a contract to Pluritec Consultants to conduct studies in the Jackson Bay area. Some members of the company visited the island in June after which "a work programme was elaborated and approved while the field work effectively took place from August 18 to August 28."[48] These events suggest that Panarctic had received a verbal approval in principle long in advance of the written approval. If so, it is possible that it had also been given guidelines for the studies in the same manner. If so it would be interesting to know if such an informal process often prevails in approvals for Arctic drilling. If there was no advance approval for drilling Jackson Bay G-16 in early 1975, it seems strange that Panarctic would have given a contract for baseline studies which encompassed only ten days of field work. Also, there is some uncertainty concerning the rig which was to have drilled the Jackson Bay well, and the timing of the operation. Separate reports on the plans for the East Drake and the Jackson Bay wells appeared in *Oil and Gas Journal* and *Oilweek*. They indicated that both wells were to have been drilled by the same rig, Commonwealth Hi-Tower Arctic Joint Venture's Rig Z (Comm-Hi Rig Z).[49] But if that was the case how could the timing for the two operations have been synchronized?

Time is one of the elements in another question about Jackson Bay: why was the well not drilled? was it a matter of lack of time to conduct two operations with Comm-Hi Rig Z? or because the Arctic Waters Oil and Gas Advisory Committee would not accept the Pluritec environmental reports or because of public pressure against offshore operations from ice islands? If it was the plan to use Comm-Hi Rig Z for both offshore operations, it is evident from the record that something went awry with the timing of the East Drake operation.[50] The ice platform for the well was completed in January but the well was not spudded until early in March and not completed until sometime in April, possibly as late as mid-month. This meant that Panarctic did not even come close to being able to drill a second well with the Comm-Hi Rig because the approval in principle for Jackson Bay had tentatively stipulated 7 May as the date to terminate drilling operations.

It would be gratifying if political pressures from the public sector had delayed the issuance of a drilling authority for the wildcat well at Jackson Bay. However, this does not seem to be realistic, in spite of the fact that it is promoted by *Oilweek*. Two articles in the same issue of the magazine stated public pressure as the reason for the delay in drilling at Jackson Bay: "A hot Panarctic prospect . . . in the Jackson Bay area was shelved by environmentalists. It was a proposed ice platform similar to Hecla, but the wildcat proposition was turned down flat."[51] Also:

Panarctic has run into considerable environmental flack due to the ice platforms. Wildcats have been ruled out, causing the demise of earlier Jackson Bay plans. Douglas H. Pimlott, Arctic Resources Committee consultant came out with a particularly paranoid article in Nature Canada suggesting DIAND was in league with oil companies to despoil the north. In the same issue, Theodore Mosquin editorialized that "the federal government has become prisoner of its own prior commitments to petroleum and pipeline development in the north and is apparently willing to take terrifying risks to fulfill those commitments."[52]

It is not known whether the Arctic Waters Oil and Gas Advisory Committee made a case against accepting the Pluritec report because it did not satisfy the conditions of the approval in principle for baseline studies and an environmental impact assessment. If it did, its actions would certainly have been justified because the report was quite inadequate, for several reasons. One was limited time: only one field season was available for field studies and only ten days of it were spent on field studies. A second problem appears to have been associated with lack of experience on the part of the scientists who conducted the study. Several elements of the study indicated that they lacked experience in the Arctic and probably with marine environments as well. For example, the report indicates that the nets which were set to sample the fish population were set at the surface just below the ice; however, most High Arctic fish are small and demersal (bottom dwellers) so could not be captured by nets set in such a way. Similarly, the survey of water quality was superficial and provided only limited information on the structure of the water column.

Still another weakness of the report was the relative emphasis placed on the study of aquatic and terrestrial environments. The drilling site is about five miles (8 km) offshore; presumably, the effects of an oil spill would be primarily on the aquatic environment and on intertidal areas and beaches. However, almost 40% of the report is devoted to reporting on the study of tundra areas which are unlikely to be affected directly by an oil spill. If adequate time had been available for field studies this emphasis might have been warranted but under the circumstances, it was not.

The faunal section of the report (Section 5) dwelt entirely on a generalized review of the potential effects of oil spills and did not provide any information on the occurrence of animals in the area. In terms of the baseline data which the report was supposed to provide, it was an important omission because the field data were very sketchy. The list of fish, birds, and mammals observed during the study included one species of fish, five of birds, and four of mammals. The report did not contain any data on these three classes of animals which could be used in the formulation of an environmental impact assessment of the effect of an oil spill.

The conclusions of the report are written after the manner of most consultants' reports which are contracted for by industry, that is, potential hazards are minimized. For example, it accepts that much of the oil from a blowout would be concentrated along a 20-mile stretch of the southwest side of Ellef Ringnes Island. Section 3 of the report concluded, "Considering the nature of the beaches and the limited amplitude of the tides, the eventual polluted zone would not exceed 30 feet in the sections along the sandy or gravelly shore. On the other hand, at the mouth of certain rivers where there are deposits of silty sediments, the polluted zone could extend inland as much as 700 feet, in exceptional cases." But one questions, with some amazement, how scientists could draw such a specific conclusion when no studies were conducted on currents in the Bay. The report contained a reference and bibliography section of 31 pages. But it contains a great deal of padding. For example, a large number of the citations, possibly as many as 50%, are not actually referred to in the text. To sum up, although the Pluritec environmental report on Jackson Bay undoubtedly cost much more to produce than the Slaney report on Offshore Hecla, it still provided remarkably little information of a relevant nature.

The point is that political pressures resulting from environmental organizations were probably not a primary factor in delaying the drilling of a wildcat well at Jackson Bay. Neither does it seem likely that adverse recommendations from the Arctic Waters Oil and Gas Advisory Committee played a primary role in delaying the project. On 3 March, *Oilweek* announced the cancellation of the Jackson Bay project; at that time the Committee had not yet submitted its recommendations on the Pluritec report to DINA. More likely, the project was delayed for a number of reasons including the breakdown of the drilling schedule at East Drake, the long delay in processing Panarctic's application for a drilling authority, and the fact that the Pluritec report was not submitted to the AWOGA Committee for review until February.

There are elements of uncertainty about the East Drake and the Jackson Bay projects which warrant public clarification before a drilling authority is issued for Jackson Bay for 1976. What were the nature and extent of the problems which were encountered at East Drake? what did they mean in terms of hazards and risks associated with the drilling of wells from ice platforms? did Panarctic meet the terms and conditions of the approval in principle and of the drilling authority in spite of the problems? If not, what ones were violated? what action did the Oil and Minerals Division of DINA take on any violations which occurred? Regarding Jackson Bay itself there is the matter of the inadequacy of the Pluritec report both as a baseline document and as an assessment of potential environmental impact. What has been done about it? has it been accepted as meeting the conditions stipulated in the approval in principle?

The Future: Oil, Gas, and the Environment

It is clear from statements made by Panarctic officials, and from the application made for the Jackson Bay well, that Panarctic plans to proceed with a programme of wildcat drilling from ice islands. Although the original proposal was made for "A system for drilling expendable delineation wells," the success of Offshore Hecla and East Drake has built up strong pressures for the approval of future wildcat operations which will be difficult to resist. No significant progress has been made towards determining the potential environmental impact of these operations. There is very little baseline information on the aquatic environment of the Sverdrup Basin and virtually nothing has been done to add to it since approval was given for the first offshore well in 1973. It is unrealistic to hire consultants to prepare environmental impact assessments as long as this situation exists.

The inadequacy of the Slaney and the Pluritec environmental assessments underscores existing questions about the present arrangements for assessments. Panarctic undoubtedly established the terms of reference and the cost limits for the Slaney and Pluritec contracts. Considering the paucity of background information on the environment of the Basin, it seems incredible that reputable consultants would have accepted contracts to produce even an initial assessment of environmental effects in a matter of days in one instance or with only ten days of field investigation. An important question for the future is whether or not the situation will be improved by the introduction of the environmental review process recently announced by DOE. The process appears to have many weaknesses[53]; considering existing jurisdictional arrangements and controls[54] it does not seem likely that regulatory processes will be greatly strengthened by the introduction of a review process.

It also seems unlikely that the companies operating in the High Arctic will invest much money in a programme of environmental research or on the development of technology which can cope adequately with the Arctic environment. Panarctic has frequently stressed the importance of spending most of its budget on exploration.[55] In addition, it appears that the company is having an increasingly difficult time financing its drilling programmes.[56] As a result, there will undoubtedly be increasing emphasis on getting maximum exploration benefits from available capital. Financial pressures might ease dramatically if enough gas were discovered to make the polar pipeline economically feasible. But until that happens, Panarctic is bound to push its present exploration capability to the limit.

Although the petroleum press called the Hecla and East Drake operations auspiciously successful there are a number of constraints that may limit the future application of the ice-island technique. The length of time required to prepare the drilling site results in a compressed drilling season and will limit the number and depth of offshore wells. At Hecla the flooding of the ice platform began in late November and continued until 3 February; drilling began on 3 March.[57] The record for East Drake was not as good. Complications developed and it was well into March before the problems were solved and the drilling operation really got underway. Since the outer limit of the drilling season is about 1 May, the effective drilling season appears to be about two months long.

Ice movement is another limiting factor. Since the drilling rig can tolerate only a limited degree of lateral movement (about 5% of the water depth at the drill site), future drilling locations will be confined to areas of relatively low ice movement. And while movement of the ice platform was within tolerable limits at the Hecla site, little precise data are available for other parts of the Arctic Basin.

In environmental terms the single most important limiting factor in Panarctic's ice-platform system is the length of time that would be required to drill a relief well in the case of a blowout. The wild well relief contingency plan submitted by the company stated quite flatly that it would require almost a full year to construct and drill from a second ice island to close off a blowout. In 1975 a second ice platform was constructed at East Drake from which another rig, which was drilling on Lougheed Island, was to have operated in case of a blowout. But, because of the date when drilling got underway, and of delays during the operation, it would have been impossible to use it if a blowout had occurred. In fact, the construction of the platform was a good public relations gesture but, in 1975, it was meaningless for purposes of environmental protection.

The use of ice islands for drilling platforms should not be the end of the technological road for drilling offshore in the Arctic Islands. Panarctic has acknowledged that other, perhaps safer, drilling systems will be available within a few years. It does not, however, seem to be doing very much to speed their development. The chief advantage of the present technique is that it is cheap — about $2 million per well — and can be used immediately. But, is it reasonable to permit the development of a full-scale exploratory drilling programme when no real action could be taken to control a blowout for a year? should approval for exploratory drilling not at least be dependent on the availability of equipment which has a much wider, seasonal capability to operate in Arctic offshore areas?

At the Northern Canada Offshore Drilling Meeting, L.G. Nicholls presented a paper entitled *The Arctic Drilling Systems (ADS).* He described the concept as "a large, totally integrated, modern drill barge with two unique and distinguishing features — an air cushion hovering system and an ice melt positioning system." He also described how the system would be used:

In the offshore regions of the Arctic Islands we have projected, again with operator guidance, that in most areas the ADS will be capable of drilling an average of 6 to 7 months per year. We do not expect to be able to drill in the summer "open water" season of the Queen Elizabeth Islands because of the massive shifting of ice and the extremely short season. It is difficult to give accurate seasonal drilling dates for the Arctic Islands because they will vary greatly with the geographic location. Generally, we would expect to be able to start drilling earlier and continue later in the western half of the Islands than in the eastern half. There may be as much as 1-1/2 months difference in the total seasons' length for these two regions. Recognizing these variables we project a typical minimum drilling season would start 1 January and end 1 July, but there are many areas where up to 10 months utilization looks feasible.[58]

He also stated, "To date, the companies which have shown the greatest interest in the ADS are Sunoco E & P Ltd. and Imperial Oil Limited. The ADS is, of course, available to any company with drilling requirements in areas where this system is suitable." No mention is made of Panarctic, a company which, perhaps most of all, has drilling requirements where such a system could be used.

The system, now called the Air Cushion Arctic Drilling System (ACADS), is said to be developed to the stage where construction can begin "as soon as a sufficiently large drilling programme justifies the expense"[59] or, one might add, as soon as the Government of Canada places a reasonable priority on the protection of the Arctic environment and insists that drilling for oil and gas will only be permitted when systems are in use which can adequately cope with Arctic conditions at any season of the year.

References and Notes

1 Statement made in a letter dated 10 September 1973 from Mr. H.J. Strain to Dr. H.W. Woodward of DINA.

2 *Oilweek, Panarctic exploration set at full $65 million over next 12 months,* 20 November 1972.

3 *Daily Oil Bulletin, Panarctic records gas discovery at ice island probe, proves advanced exploration techniques,* 2 April 1974.

4 Baudais, D.J., Masterson, D.M., and J.S. Watts, *A system for offshore drilling in the Arctic Islands, Journal of Canadian Petroleum,* July-September 1974; also Nowell, R., *Panarctic drills offshore well from Arctic ice platform, Canadian Petroleum,* May 1974.

Self-contained Air Cushion Arctic Drilling System (ACADS) concept is one possible solution to Arctic environmental problems. Drawing shows unit under tow across sea ice by largewheeled vehicles. Smaller ACT is bringing supplementary supplies.

5 This account of the environment of the region draws heavily on published papers and reports. Some of these are extensively quoted. In some places some licence was taken in changing the order of presentation and the paragraph structure.

6 Taylor, Andrew, *Our polar islands — the Queen Elizabeths, Canadian Geographical Journal*, June 1956.

7 *The climate of the Canadian Arctic, Canada Year Book* (Ottawa: Dominion Bureau of Statistics, 1967).

8 See n. 7.

9 See n. 7.

10 See n. 7.

11 See n. 7.

12 See n. 7.

13 See n. 7.

14 See n. 7.

15 Slater J.E., Ronhovde, A.G., and Van Allen, L.C., *Arctic environment and resources* (Montreal: Arctic Institute of North America, 1971).

16 Collin, A.E., *The waters of the Canadian Arctic Archipelago, Proceedings of the Arctic Basin Symposium* (Montreal: Arctic Institute of North America, 1962).

17 See n. 6.

18 See n. 15.

19 Ewing, G.N., *Arctic offshore knowledge and research, A review of arctic hydrography*, minutes, Northern Canada Offshore Drilling Meeting, Ottawa, 5-6 December 1972.

20 See n. 15.

21 See n. 16.

22 Dunbar, M.J., 1970, *On the fishery potential of the sea waters of the Canadian North, Arctic Journal*, vol. 23, no. 3, 1970.

23 See n. 15.

24 MacDonald, S.D., *Report on biological investigations at Alert, N.W.T.* (Ottawa: National Museum of Canada 1950), Bulletin 128.

25 Canada, Canadian Wildlife Service, *Arctic Ecology Map Series*, "Critical Wildlife Areas" (Ottawa: Canadian Wildlife Service, 1972).

26 See n. 25.

27 Tener, J.S., *Queen Elizabeth Islands game survey, 1961* (Ottawa: Canadian Wildlife Service 1963), Occasional Paper No. 4; see n. 31.

28 See n. 25.

29 Heise, H., *First Arctic offshore tests to be drilled on Kristoffer Ice, Oilweek*, 12 March 1973.

30 Letter from Strain to Woodward, see n. 1.

31 Baudais et al., see n. 4.

32 Panarctic Oils Ltd., *Six year review and the future*, 6th Annual Report 1973.

33 Panarctic Oils Ltd., *A system for drilling expendable delineation wells for the offshore Arctic islands,* prepared by Westburne Engineering, September 1973.

34 Included in the appendices of Westburne report, see n. 33.

35 Panarctic Oils Ltd., supplement I, *A system for drilling expendable delineation wells for the offshore Arctic islands,* September 1973.

36 F.F. Slaney and Co. Ltd., Vancouver, September 1973.

37 Letter addressed to Mr. H.J. Strain, Panarctic Oils Ltd., from R. Webb, Chief Ecologist of F.F. Slaney Co.

38 Minutes of the Department of the Environment Management Committee meeting, 6 September 1973, item 34.6.

39 In fact the sand was hit at 2695 feet so the estimate was quite accurate. See McGhee, E., *Panarctic pioneers floating-ice drilling, Oil and Gas Journal,* April 1974.

40 Strain, H.J., *Drilling in the High Arctic, Petroleum Engineer,* January 1972.

41 Canada, DOE, Environment Protection Service, *Recommendations on waste and hazardous materials — Management on Panarctic Hecla O-62* (Ottawa: DOE, 1973).

42 *Arctic Islands to get second offshore test, Oil and Gas Journal,* 3 March 1975.

43 *Panarctic plans second offshore platform test, Oilweek,* 16 December 1974.

44 *Eight-mile step-out extends Drake Point gasfield, Oil and Gas Journal,* 21 April 1975.

45 Letter to Mr. H.J. Strain, Panarctic Oils Ltd., 4 October 1974 from Mr. F.J. Joyce, Director, Northern Natural Resources and Environment Branch, DINA (copy of letter and attachment, see n. 22, provided by the Department).

46 Attachment 1, "Approval in principle (for drilling a well from an ice platform at Jackson Bay S-16)." The attachment stipulated specific terms of approval in principle which was granted by letter, see n. 21.

47 Slightly condensed from section 1.3 of the Attachment, see n. 22.

48 Pluritec Consultants, *Baseline environmental and resource data on flora, fauna, sediments and soils, Jackson Bay, Ellef Ringnes Island, N.W.T.,* Contract No. 4085 for Panarctic Oils Ltd., 1975.

49 *Arctic Islands to get second offshore test, Oil and Gas Journal,* 3 March 1975, and *Eight-mile step-out extends Drake Point gas field* and *Panarctic plans second offshore platform test, Oilweek,* 16 December 1974.

50 We received information from an authoritative source that Panarctic encountered several problems in drilling the East Drake well; one of these resulted in a segment of the casing being pulled up and damage to the marine riser and BOP stack. This resulted in some of the delay to the operation but it does not account for the long delay between the construction of the ice platform and the commencement of drilling.

51 *Mackenzie Delta hums with activity despite exploration and political setbacks, Oilweek,* 3 March 1975.

52 Pace of Arctic Islands exploration hinges on discoveries and regulations, *Oilweek,* 3 March 1975.

53 Canadian Arctic Resources Committee, *The impact policy — empty rhetoric, Northern Perspectives,* vol. 3 no. 3, 1975.

54 Pimlott, D.H., *Delta gas: time and environmental consideration,* in *Gas from the Mackenzie Delta, Now or Later?* (Ottawa: Canadian Arctic Resources Committee, 1974).

55 *Oilweek,* 20 November 1972. See n. 2.

56 *Pace of Arctic Islands exploration hinges on discoveries and regulations, Oilweek,* 3 March 1975, and *Fund raising slow for Panarctic 1975-76 plan. Oil and Gas Journal,* 12 May 1975.

57 Baudais et al., see n. 4.

58 L.G. Nicholls, *Arctic drilling systems (ADS),* Northern Canada Offshore Drilling Meeting, DINA, December 1972.

59 Jones, G.H., *Arctic poses extreme technical challenges, Petroleum Engineer,* January 1975.

Chapter 6

Drilling in Hudson Bay

Hudson Bay is not considered as a major exploration play by the oil industry. Although more than 38 million acres offshore are under exploratory permit, the sedimentary basins beneath the Bay are still largely a mystery to petroleum geologists, even though the first Canadian offshore well in Arctic or sub-Arctic waters was drilled in Hudson Bay — an incident both industry and government would prefer to forget, and one that offers a sharp contrast to confident expressions of reassurance about large-scale offshore operations in the North.[1] The well was drilled by Aquitaine Co. of Canada in 1969 and ended in near disaster when the drilling rig "lost hole" and sustained $400,000 damage in a severe storm. Aquitaine returned to the Bay in 1974 with a new semi-submersible drilling system — the first of its type to be used in Arctic waters — to plug the unfinished 1969 hole and drill two new wildcat wells.

Although a tôtal of 625 exploratory permits are held in the Bay, more than half of the total acreage under permit is held by only four companies: Atlantic Richfield (12 million acres), Aquitaine (4.3 million acres), Sunlite Land Ltd. (3.6 million acres), and Seibens Oil and Gas (1.8 million acres).[2] Acquisition of acreage was at its peak in 1968, and since then has declined every year as the companies have diverted their exploration funds to regions with more immediate promise.

Although the Hudson Bay basin, which includes both the offshore areas and the adjacent onshore Hudson Bay lowlands, has not been extensively explored (a seismic survey of some 5000 miles was scheduled for the 1974 summer season), a recent review of exploration prospects in *Oilweek* offered this appraisal: "The hydrocarbon prospects are considered excellent by many and 'dicey' by some. It is a new ball-game and information is scanty. Best and most positive data are held by land-owners and they're not particularly anxious to develop increased competition through disclosure at this time."[3] Despite the paucity of published information, the magazine's editors managed an upbeat conclusion. The Bay, they said, "shakes up as a pretty potent energy cocktail." Potent or not, Hudson Bay poses some formidable challenges to drillers.

The Hudson Bay Environment

In spite of its "southern" location, the Hudson Bay region is remote and logistic support is probably poorer than in any other part of the Arctic. The physical environment of the Bay, while not as harsh as the Beaufort Sea, does pose major risks to any drilling operation, as Aquitaine discovered five years ago. Reliable meteorological and oceanographic forecasts are non-existent. In the middle of the Bay, drillers must also contend with shifting summer pack ice. Aquitaine's exposure in 1969 to these offshore conditions led to a re-evaluation by the company and by government officials of the type of drilling equipment needed for future work in the Bay.

Hudson Bay, with James Bay, has been described as forming an immense inland sea some 300,000 square miles in area, nearly ten times as large as Lake Superior . . . it is comparable in area to the North Sea and the western part of the Baltic. The shipping route from the eastern entrance of Hudson Strait to Churchill is approximately the same length as the European route from Helsinki to Aberdeen or the Orkney Islands.

The contrast between the Arctic and sub-Arctic climate of Hudson Bay and the areas of comparable latitude in Europe is well illustrated by the differences in ice conditions along these two water routes and the influence which they have on neighbouring land areas. The North Sea and its connections with the Baltic are for the most part ice-free throughout the year, owing to heat transfer by winds and ocean currents from the Atlantic Ocean. The waters of Hudson Bay, on the other hand, receive no ameliorating influences such as come to the waters of western Europe from the Gulf Stream and the relatively warm eastern Atlantic. As a result, they are blanketed throughout much of the year by a cover of ice which, in winter, acts as an extension of the snow-covered northern land masses, permitting Arctic winds with their frigid temperatures to sweep unmodified over its relatively level frozen surface far southward into Ontario and Quebec.

The outstanding characteristic of the ice of Hudson Bay is its variability. Although a virtually complete ice cover spreads over the entire area each winter and disappears in the course of the following summer, the pattern of its formation and break-up differs widely from year to year and from one locality to another.[4]

The three wells which have been drilled in Hudson Bay have all been south of the 60th parallel. From a latitudinal view, they might reasonably be referred to as being in the "north" but not in the Arctic or sub-Arctic. But that is a technicality which does not warrant application in the case of Hudson Bay. A multitude of other characteristics — the nature of the vegetation and the presence of permafrost in the surrounding land mass, the nature of the fauna and flora of the Bay, and the annual temperature and ice regimes — show that "Hudson Bay and its environs have a typically Arctic climate."[5]

Ice begins to form in the northern parts of the Bay (approximately 65°N) in early October; by the end of the month the rivers are normally frozen and a narrow bank of shore-fast ice closes off the smaller bays. The southern bays and rivers are generally sealed off by mid-November:

Although the ice coverage of Hudson Bay is usually at its maximum from January to April it is never 100 percent complete. The shearing effects of winds, tides and currents produce stresses that fracture the ice, separating floes and even large solidly-frozen ice fields from each other and either driving them apart, thus creating stretches of open water, or causing them to override so that they are crushed and broken into a maze of ridges, mounds and hummocks . . . lanes of open water known as leads develop and may vary in length from a few yards to several miles, and in width from a few feet to channels a mile or more across. The most outstanding of these is the shore lead or flaw lead between the fast ice of the coast and the free-floating ice of the deeper waters.[6]

The surface of the Bay is 95-99% covered with thick pack ice from late December through April. The ice gradually decreases from 30% to 10% coverage by July. The Bay is generally ice-free from August through October except for shore ice which begins to form in late October.

The average depth of water over the main area of Hudson Bay is about 100 metres (340 feet) although depths range to about 230 metres (750 feet) in the north-central region where depths increase progressively from the shoreline outward. The deepest areas of the Bay however occur in troughs; one west of Ottawa Islands is 300 metres (1017 feet) deep and the other is adjacent to Digges Island near Cape Wolstenholme with a depth of 550 metres (1865 feet). Exclusive of these exceptionally deep troughs, the general bathymetric configuration of Hudson Bay is saucer-shaped.[7]

The water circulation in the Bay is counter-clockwise. This prevailing pattern of the currents, combined with tidal action and strong winds, results in complex patterns of water movement which break up or compress ice masses, move them around the Bay, open and close leads in an indescribable array of patterns.[8]

It would appear that Hudson Bay is primarily Arctic in its invertebrate fauna . . . Hudson Bay would appear to be neither rich nor exceptionally poor in the diversity of the invertebrate animals and in the quality of animals it supports. It would seem to belong in an intermediate position to the rich North Atlantic sub-Arctic on the one hand and to the deficient Arctic Basin on the other . . . the fish fauna is largely typical of Arctic regions and few species are present. Abundance of all species is low, growth rate is slow and productivity is depressed. As a consequence of this limited potential, commercial exploitation has been possible on a marginal basis on only a few species. Attention to the fishes has been minor and relatively few reports have been published.[9]

Thirty-seven species of marine and anadromous fish have been recorded for Hudson and James Bay: "Nearly all of these species are small, obscure, bottom-dwelling forms occurring in low abundance and devoid of any apparent value in the economy of man."[10] Since the water of the Bay is of polar origin, it has an Arctic/sub-Arctic component of marine mammals, which includes the ringed seal, bearded seal, harp seal, harbour seal, walrus, beluga or white whale, the killer whale, and the bowhead or Greenland whale.[11]

The Bay is an important migration route, staging and nesting area for many species of birds. Digges Island, near Cape Wolstenholme, has a nesting colony of thick-billed murres which has an estimated nesting population of 2-3 million birds.[12] The birds of this colony have been utilized for food by the Inuit of the region for a very long time. Several other species of birds also constitute important resources for the indigenous people as do the harbour, ringed, and bearded seals, walrus, beluga, and Arctic char.

The First Offshore Well in the North

The official name for Aquitaine's first well was "Aquitaine et al Hudson-Walrus 71." Application to drill the well was submitted by the company in the spring of 1969 and approved by officials of the Resource Management and Conservation Branch of the Department of Energy, Mines and Resources (EMR).[13] The drilling vessel chosen by Aquitaine for the Walrus operation was the Wodeco II, a barge-type rig which had spent most of its drilling life off the California coast. The Wodeco II had also drilled in Cook Inlet off the coast of southern Alaska, and its crew was regarded as experienced in Arctic operations, an important factor in EMR's decision to approve Aquitaine's application, although subsequent events revealed that conditions in Cook Inlet are in no way comparable to the more difficult conditions facing operators in Hudson Bay. (Despite these differences, the DINA offshore drilling memorandum to Cabinet three years later blithely attempted to compare conditions in Cook Inlet to conditions in the Beaufort Sea which are even harsher than those of Hudson Bay.) Early in the summer of 1969, the Wodeco II rig was towed down the California coast, through the Panama Canal, up the east coast of North America and into Hudson Bay via Hudson Strait. Normally, the ice-free "drilling window" in the Bay is about three months from late July through most of October, depending primarily on ice conditions in Hudson Strait which is the only access route to the Bay. With icebreaker support, it is estimated that the drilling season could be extended to about 120 days.

Aquitaine did not get a good start in 1969. Although the company had hoped to begin drilling earlier, the Walrus well was not spudded until 7 August because of slow ice break-up at the well location (58N, 87W). The operation ended prematurely on 16 October at which time the well had been drilled to a depth of only 3926 feet. In a press release shortly after ending operations, the company said that the "Decision to suspend the well was made in anticipation of the seasonal build-up of ice in Hudson Strait which would impede removal of equipment from the Bay . . . . Late departure of ice from the location prevented spudding before August 7 and wave and weather conditions were causes of difficulty throughout the drilling operation."[14]

At best, this was a rather bland description of the difficulties encountered during the operation. In an official abandonment report submitted to EMR, Aquitaine gave quite a different account of events which led to suspension of drilling: "Due to very severe weather conditions, the connection between the drilling vessel and the well was lost October 16. Because of severe damage to the equipment and the necessity to leave the Bay before the Straits were closed by ice, it was impossible to resume operations and plug the well in a conventional manner."[15] The storm that ended the Walrus operation was accompanied by waves of up to nine metres and winds that gusted up to 70 knots, not unlike the storm conditions that are prevalent on the Beaufort Sea during the late summer and early autumn. The storm struck without any warning or advance forecast. As it gained in intensity, all unnecessary personnel were evacuated to Churchill. A subsequent memorandum submitted to EMR and containing excerpts from the log of the project manager for the drilling contractor, Western Offshore Drilling and Exploration Co., provides a vivid picture of events at the height of the storm: "1200 hrs: Winds changing to NW and gusting to 60 knots. Travelling blockstore guidance system loose. Blocks were dropped to rig floor. 28 stands drill pipe broke loose from lashing, tearing out racking fingers and boards, breaking and bending girders and braces. Noise was terrible. Rig floor has suffered much damage. Vessel motion still severe."[16]

According to an EMR official it was believed at the time of the application that the company's previous geophysical work in the Bay would have given Aquitaine adequate foreknowledge of some of these problems. But, *Oilweek* noted later, "at that time [1969] little was known about waves, currents and weather patterns in the area."[17]

In fact, the Wodeco II rig had been plagued with problems from the beginning of the operation, one of which was that the wave period of "this big and sophisticated drilling vessel" happened to be the same as that prevalent in Hudson Bay. This caused "strong and bothersome" movements.[18, 19]

In retrospect, the Walrus incident appears to have been a case of proceeding by trial and error. Neither the company nor EMR had any real appreciation of physical conditions in the Bay. At the time, the staff of the Energy Conservation Branch responsible for granting the drilling authority was far too small to conduct a thorough assessment, and other government agencies were given little if any opportunity to review Aquitaine's plans. According to departmental procedure at the time, all offshore proposals were to have been sent to a number of federal agencies for review and comment, including the Department of National Defence, the Ministry of Transport, the National Research Council, and the Department of Fisheries and Forestry (the forerunner of the Department of Environment). It appears, however, that Aquitaine's proposal was never forwarded to Fisheries and Forestry. A letter from the Resource Management and Conservation Branch to Aquitaine requesting that the company send a copy of its proposal to Fisheries and Forestry appears in the files with the comment "never sent".

In any case, the drilling application submitted by Aquitaine did not include supporting environmental data, documentation of environmental research, or contingency plans in the event of a blowout. No federal studies on possible environmental effects were undertaken either prior to or concurrent with the drilling programme. Had an accident occurred late in the drilling season, it would have been impossible to drill a relief well until the following season when equipment could again be moved through Hudson Strait. Had the Walrus well been "kicking" (actively encountering an oil pool) at the time of the 16 October storm, the possibility of a blowout would have been almost certain. The drill did encounter some trace findings of hydrocarbons, but there has been no evidence of oil seeps during the five years the well was left unplugged.[20]

Resumption of Drilling, 1974

Aquitaine planned to return to the Bay in order to plug the Walrus well and to drill two new wildcat wells, and 1973 was to be an intensive year for both onshore and offshore activities including two offshore wells and intensive marine seismic work.[21] The Pentagone P-82, very different from the Wodeco II, was to drill the wells.

By any standard, semi-submersibles of the Pentagone type are impressive machines, "the Cadillacs of the offshore drilling business," in the words of one official. The P-82 is 325 feet long, 338 feet wide, and 134 feet high to its upper deck. The rig is ice-reinforced and able to break new ice up to one foot deep, and able to withstand contact with small ice floes. Aquitaine had originally ordered the rig in 1970 in anticipation of drilling during the 1973 season, but because of an EMR policy that no rig be allowed to drill its inaugural well in waters under EMR jurisdiction, it was sent to the North Sea until the start of the 1974 summer drilling season. In lieu of drilling, an extensive seismic programme was conducted in 1973 by Kenting Exploration. It was supported by about 10 companies and data were obtained on 1900 miles of reflection seismic.[22] An additional 5000 miles of seismic data were later added during the time the subsequent Aquitaine wells were being drilled.

In spite of the apparent lessons of the 1969 drilling programme, the administrative machinery for assessment seemed no more able to keep pace with events in 1974 than it had five years earlier. On 22 May 1974 Aquitaine officials met with officials of EMR to outline the final details of their application for a drilling permit under the Oil and Gas Conservation Act. By the time of this meeting, however, there was no doubt that Aquitaine would indeed be drilling in Hudson Bay in the summer of 1974 since programme approval (equivalent to approval in principle) had already been granted at an earlier date. At the 22 May meeting Aquitaine outlined the final details of its plan to use a semi-submersible rig, the Pentagone P-82, in the 1974 drilling programme, the first use of semi-submersible technology in Arctic waters.

The rig was to be towed off its location in the North Sea early in June, arrive at Cape Chidley early in July and receive ice-breaker support through the Strait and into Hudson Bay where it was scheduled to begin operations in late July. The current drilling programme was a more costly effort than the earlier Walrus venture. For a price of $12 million, Aquitaine hoped to drill two wells: Narwhal No. 1, 185 miles south-west of the original Walrus well, and Polar Bear, 15 miles to the east of Walrus. Working interest in the programme was divided among four firms:

Aquitaine had 26.3% interest in the venture, Shell Canada Ltd. 60.5%, Petrofina Canada Ltd. 6.6%, and Sogepet, 6.6%.[23]

By the time the Department of Environment became involved in the assessment process, the drilling programme had been set in all but the smallest of details. Confronted with this *fait accompli,* the department once again found itself on the defensive. It requested EMR to have Aquitaine submit a contingency plan and an environmental assessment.[24] Aquitaine met the requirement, but it was clear that the company did not regard environmental assessment as a crucial or deciding factor in its application for a drilling permit. As though to underscore its indifference to the whole matter of environmental assessment, Aquitaine's submission to DOE did not include a contingency plan to deal with an oil spill, nor any provisions for securing a rig to drill a relief well in the event of a blowout, nor any indication of how quickly a blowout might be dealt with. Neither did the assessment offer any clues as to how adverse weather conditions might affect any relief operations. In spite of the Walrus episode in 1969, Aquitaine's assessment omitted any reference to the factors that led to the improper abandonment of the Walrus well, although one of the major objectives of the 1974 programme was to complete that hole in a proper fashion. Perhaps most amazing was the company's assertion that winds in the Bay are "light but steady," when records from the Churchill weather station indicate that in September, the station experiences the highest mean wind speed of any weather station in Canada.[25]

The DOE review of Aquitaine's submission added this comment on the inherent risks from wind and wave action:

In the offshore, the possible effects of the physical environment on the drilling system constitute one of the principal risks in such an enterprise. DOE considers these risks to the operation as potential threats to the surrounding environment. Wind and wave forecasting services in Hudson Bay are non-existent. If various aspects of the operation depend on advance notice of critical values of wave heights, the lack of such a service may pose an unacceptable risk. Similarly weather constraints related to contingency actions should be considered.[26]

Aquitaine's drilling programme had been five years in the planning, but officials in Environment Canada had time to give the Aquitaine assessment only a cursory review and prepare a short statement reflecting the department's concerns about the drilling venture before events began closing in. By late June, only a few weeks after DOE had received the assessment, the Pentagone rig was ready to move from the North Sea to Hudson Bay to begin drilling. To delay the start of drilling would cost Aquitaine and its partners millions of dollars, a predicament that would be unacceptable to both the company and EMR.

On 5 July, officials of EMR and DOE met to discuss the environmental implications of the programme. EMR agreed to incorporate four suggested conditions in the drilling permit issued to Aquitaine, including a request for clarification of contingency procedures, a requirement for studies on the effects of drilling on aquatic life, a request for data on weather conditions, and restrictions on the discharge of liquid wastes from the drilling rig. In the light of DOE's statement of its concerns about the drilling programme, the four conditions are indeed a modest response to the Aquitaine programme.

Environment Canada has a primary concern for the well being of the living natural resources of Hudson Bay and the ecological systems that support these resources.

Because of their sensitivity and slow rate of recovery extreme care should be taken in any activity which will impinge on northern ecosystems.

Phytoplankton, zooplankton and invertebrate populations (which occur to depths of 100-180 m) provide food supply for benthic and pelagic (including anadromous) fish populations.

Huge populations of sea birds occur in the bay. Leads in sea ice are occupied by seabirds throughout the winter. During the fall migration period large numbers of seabirds, waterfowl and shorebirds concentrate along the tidal wrack to rest and feed. Species such as murres and eiders are dependent on the sea during all seasons of the year.

Marine mammals occur throughout the Bay. Pods of whales have been observed in the Bay far from shore while polar bears range up to 250 miles from shore to hunt seals which in turn utilize fish as their only source of food.

The principal concern of DOE is with a major oil spill or continuous leak of oil due to unforeseen events at the site, e.g. release of oil due to improper or inadequate well completion, damage by ice, structural (geological) failure or blowout. The magnitude and duration of the effects of a major oil spill are expected to be beyond compensation due to the sensitivity and slow rate of recovery of northern ecosystems. An oil spill or oil blowout could contaminate vast areas of shoreline in Hudson Bay, and because of the emulsification of oil in cold water, large areas of bottom could be disrupted. Oil on the surface could have extremely adverse effects on the sea bird populations by either directly killing them by coating with oil and destroying the insulating properties of the feathers or by ruining the birds' food supply and eliminating them by starvation. Oil spills could be deleterious to the marine mammals and oil-in-water emulsions could prove lethal to fish and fish food organisms. Of major importance are the ringed seal, walrus and beluga. The result of such damage on marine resources could affect the resource base available for native peoples' use.[27]

But the statement of concerns and ideals does not seem to have been backed up by a reasonable show of determination by DOE. The Department had five years' lead time after the first operation to establish its position and make its requirements known, and there were reminders along the way that indicated that plans for additional work in the Bay were going forward.[28] The opportunities for stating a strong case included participation at the Northern Canada Offshore Drilling Meeting in December 1972, where DOE played a very weak role.[29] It is possible that the department's interest had not really been stirred until July 1973, when Cabinet considered and gave approval in principle to offshore drilling in the Beaufort Sea. At any rate, it was 9 July before an official at the Assistant Deputy Minister level stated his case to his opposite number in EMR, just 13 days before *Oilweek* reported that the Pentagone rig was holding up in Baffin Strait waiting for the ice to clear from the Bay.[30] On 5 August it was reported that the rig had anchored at the Narwhal site and was preparing to drill. It is not known whether DOE received an official EMR response to the DOE Environmental Position Paper and the conditions it proposed for the Aquitaine application.[31]

The 1974 drilling programme continued until late October. Narwhal was drilled to a total depth of 4341 feet and abandoned early in September; Pentagone was then moved to a new site to drill a third well, called Polar Bear. En route to the Polar Bear location Pentagone "went back to button down the first offshore well ever sunk in the vast grey emptiness of Hudson Bay in 1969."[32] (Walrus 71 had then existed as a continuing threat to the environment of Hudson Bay for almost six years). Polar Bear was drilled to a depth of 4500 feet by late October when Pentagone was towed back through Hudson Strait and across the Atlantic to a new location in the North Sea.

Plans for future drilling in Hudson Bay appear to be indefinite. At the end of the 1974 season the *Financial Times* ran a brief note downgrading the prospects for major discoveries of oil and gas. It pointed out that "Aquitaine cannot be all that excited because it has dropped a lot of the acreage it once held in the Bay."[33] The same week a series of articles from the drilling site in Hudson Bay gave a much more optimistic outlook: "Four well sites have been tentatively selected for the next round of tests on geological structures that are usually regarded as likely traps on land or offshore for oil and gas accumulations: Aquitaine is bullish in Hudson Bay, partly because it is the only company with any practical information on the region, and it has come nowhere near disillusionment in attempts to grasp the potential in only two seasons of active drilling."[34] At any rate Aquitaine did not go back to drill in Hudson Bay in 1975 and no definite statements have been made about its intentions for the future.

The Concerns and Viewpoints of the Native People

The native people who live on or near Hudson Bay are not likely to be disappointed if the petroleum industry decides that the Bay has a low priority compared to exploration prospects in other parts of the world. Ironically, it was DINA that informed some Inuit communities about the proposed drilling programme for the Bay in March 1974 when the communities were visited by the Regional Director of Resources of DINA, Bill Armstrong, who was on a tour of Keewatin and the eastern Arctic. He was accompanied by Fred Joyce and Dr. Maurice Ruel, Director and Assistant Director respectively of the Northern Natural Resources Branch of DINA. The Director of Resources was simply passing on information; because the drilling came under the jurisdiction of EMR, he had no direct role in the regulatory process. A member of the Inuit Land Use and Occupancy Study was working in the area at the time, and she reported to the president of Inuit Tapirisat of Canada that no one she had talked with in Chesterfield Inlet, Eskimo Point, Rankin Inlet, and Whale Cove had been consulted or informed about the exploration which was planned for that summer in the Bay.[35] On 29 April 1974 a public meeting was held in the community hall at Chesterfield Inlet to discuss exploration for petroleum and minerals in the Keewatin district. The public statement which was issued following the meeting contained this statement:

In the past few years, there has been a large influx of mineral and oil exploration in the Canadian Arctic and still more exploratory activities are coming into existence. Due to the fact that seismic programmes and offshore drilling are now being effected in and on Hudson Bay, the Inuit people feel that the danger to the wildlife and the sea animals has increased tremendously; therefore, some measures of control over the exploration companies should be imposed in consultation with the Inuit people. Specifically, the people of Chesterfield Inlet are opposed to any seismic activity and offshore drilling in and on the Hudson Bay since it would greatly affect the sea animals, not only around the immediate area of exploration but within the waters of Hudson Bay. Should there be a discovery of oil under the Bay, there is a great danger of oil spillage or leakage which may float up from the bottom of the Bay and due to sea currents may surface miles away and may not be discovered until enormous amounts of damage have already occurred. An example of oil in the sea and its effects on sea animals was experienced at Chesterfield about a year ago. There was a leak in one of the bulk fuel tanks and during the spring thaw runoff, the oil floated down into Mission Lake and via a small creek, ended out on Spurrell Bay at the mouth of Chesterfield Inlet. Throughout the summer, there were no fish in Spurrell Bay and seals could only be caught far away from their normal habitat around Spurrell Bay.[36]

On 22 May the *Winnipeg Free Press* reported that the secretary-manager of the Hamlet of Coral Harbour had written Donald Macdonald, Minister of Energy, Mines and Resources, to express the concerns of the members of the community over the exploratory drilling programme. Five specific questions were posed:[37]

1) *Have environmental studies, similar to those proposed for the Beaufort Sea area, been completed in Hudson Bay?*
2) *Does the federal Government, which has given approval in principle, have full information on the effects of an oil spill on the water coastline and animals of Hudson Bay?*
3) *Does Aquitaine know exactly what problems may be encountered in drilling operations this year and can they guarantee that no accident, similar to that at Walrus Hole in 1969, could occur?*
4) *If an accident does occur, can the federal Government guarantee that an oil spill (or blowout) can be handled without environmental damage?*
5) *Is there a second rig available in Hudson Bay which can plug a blowout hole if damage occurred to the first rig, or would an oil blowout continue all winter until the hole could be plugged after breakup?*

Macdonald's reply was the traditional sort of response by a Minister to an individual or an organization which has expressed concern about the policies of his department. He described the Pentagone P-82, Aquitaine's contingency plan, and the surveillance of the drilling operation by officials of his department, in very positive terms:

The precautionary and preventative measures we take with respect to drilling in the Canadian offshore are exhaustive in their specifications and are especially strict by world-wide standards.

With respect to your comment that you were not visited this spring by a group touring the Keewatin settlements, I have determined that this group was organized by the Department of Indian Affairs and Northern Development for the purpose of discussing plans for future commercial and industrial activity in the District. However, since the Aquitaine drilling programme was not within the District or close to any of the settlements, it was not included as a subject for discussion, and representatives from Energy, Mines and Resources and Aquitaine were, therefore, not to participate.

In response to your specific questions, my officials assure me that, as with all proposed drilling programmes, the Aquitaine programme has been carefully examined and assessed to ensure that the risk of accident is minimized. An environmental assessment has been made, including a catalogue of all the species of wildlife. It is interesting to note that a large area in the centre of the Bay, in the areas where drilling is to be carried out, is classified by some biologists as a biological marine desert. It is also estimated that, even under the worst conditions, it would take over a month for any oil spilled at the proposed drilling locations to reach the nearest shore. This means that the necessary clean up operations could be initiated and be in full operation long before the oil got close to the coast.

According to the Minister, two drillships that were working in the Labrador Sea would be available within a few days to drill a relief well if one were required. However, he neglected to say anything about what would happen if a blowout occurred late in the season when the drillships would not dare to enter Hudson Bay. The description of the problems encountered in drilling Walrus was a masterpiece of understatement: "With reference to your comments about the accident involving the Walrus well, this was really more in the nature of a mechanical failure than an accident. Although it cost the operator a great deal of money and necessitated the termination of the operation it did not involve injury to personnel or the spilling of any oil."[38]

The efforts of the Coral Harbour Settlement Council to clarify their position on offshore drilling were not very productive, but they did indicate that native people would not gain much if the Department of Energy, Mines and Resources replaced DINA as the regulator of the oil industry in the North as has sometimes been suggested. It is also evident that EMR has problems focusing on the interests of native people when multinational companies are involved.

References and Notes

1 Our information on the drilling of this well was obtained from a review of the file at the Department of Energy, Mines and Resources. One of us was permitted to read and make notes on things of interest to our investigation.

2 These figures were obtained in 1974, prior to the drilling programme.

3 *Aquitaine plans two Hudson Bay wells, Oilweek,* 1 April 1974.

4 Lardner, M.M., *The ice,* in *Science, History and Hudson Bay,* edited by C.S. Beals (Ottawa: DEMR, 1968).

5 Thompson, H.A., *The climate of Hudson Bay,* in *Science, History and Hudson Bay,* see n. 4.

6 Lardner, M.M., see n. 4.

7 Pelletier, B.R., F.J.E. Wagner, and A.C. Grant, *Marine geology,* in *Science, History and Hudson Bay,* vol. 2.

8 Lardner and Pelletier, see n. 6 and 7.

9 Grainger, E.H., *Invertebrate animals,* in *Science, History and Hudson Bay,* vol. 1.

10 Hunter, J.G., *Fish and fisheries,* in *Science, History and Hudson Bay,* vol. 1.

11 Mansfield, A.W., *Seals and walrus* and Sergeant, D.E., *Whales,* in *Science, History and Hudson Bay,* vol. 1.

12 Tuck, L.M., *The murres,* Monograph Series No. 2 (Ottawa: Canadian Wildlife Service, 1960).

13 Offshore drilling in both Hudson Bay and Hudson Strait falls within the jurisdiction of the Department of Energy, Mines and Resources (EMR). This department administers the Canada Oil and Gas Land Regulations and the Oil and Gas Production and Conservation Act in waters off the east and west coasts and in Hudson Bay and Hudson Strait. The Department of Indian and Northern Affairs administers the same legislation and has jurisdiction over offshore operations in the rest of the Arctic, including Davis Strait and Baffin Bay.

14 *Major rush planned by oil hunters for next year in Hudson Bay, Oilweek*, 4 December 1972.

15 Memo from Aquitaine to EMR.

16 Memo from Drilling Supervisor on Wodeco rig to Aquitaine and forwarded to EMR.

17 *Major rush planned by oil hunters for next year in Hudson Bay region, Oilweek*, 4 December 1972.

18 Two years later, the ill-fated Wodeco II rig was destroyed in a blowout and fire in the offshore waters of Peru. The $4.5 million disaster took seven lives. Thobe, S., *Storms are a major cause of mobile rig mishaps offshore, Offshore*, 5 June 1975.

19 *Oilweek*, see n. 5.

20 According to an official of the Resource Management and Conservation Branch of the Department of Energy, Mines and Resources.

21 *Oilweek*, see n. 15.

22 *Hudson Bay seismic data encouraging — more reflection surveys will be shot, Oilweek*, 1 April 1974.

23 *Globe and Mail, Drills set to probe off Hudson Bay shore*, 31 July 1974. Also *Aquitaine's Hudson Bay oil payoff would be proximity to markets*, 16 October 1974.

24 Department of the Environment, environmental position paper, July 1974 re *Aquitaine Company of Canada application to drill in Hudson Bay.*

25 DOE, see n. 22.

26 DOE, see n. 22.

27 DOE, see n. 22.

28 For example, the review of plans which was given in *Oilweek*, 4 December 1972, see n. 15.

29 See Chapter 2.

30 *Oilweek, Pentagone P-82 outwaits ice for drilling in Hudson Bay*, 22 July 1974.

31 Letter from Mr. K.C. Lucas, Senior Administrator, Fisheries and Marine Service, DOE to Mr. W.H. Hopper, ADM, EMR, 9 July 1974.

32 *Globe and Mail*, 16 October 1974, see n. 21.

33 *Financial Times, Poor prospect*, 21 October 1974.

34 *Globe and Mail*, 16 October 1974, see n. 21.

35 Letter from Lois Little to Tagak Curley, 19 April 1974.

36 Chesterfield Inlet, *Public statement*, 29 April 1974.

37 *Winnipeg Free Press, Hudson Bay Offshore Drilling*, 22 May 1974.

38 Letter from the Hon. Donald S. Macdonald, Minister of EMR to Mr. Mike Cluderay, Secretary Manager, Hamlet Council of Coral Harbour, 19 June 1974.

Chapter 7

Lancaster Sound and the Northwest Passage

Lancaster Sound is one of the most biologically productive offshore regions in the entire Arctic and the most recent to capture the attention of the oil industry. Even in the Arctic, where great technological risks seem commonplace, the proposal to drill in Lancaster Sound has a science-fiction air about it. Submitted to DINA in February 1974 by a firm called Norlands Petroleum Ltd., the proposal outlined plans to drill a wildcat well in waters more than 2500 feet deep — deeper than any such well ever drilled.[1] Late in the summer of 1974, DINA gave its approval in principle to the project.[2]

The approval specified that "Drilling operations shall be restricted to the ice-free periods . . . between July 15 and October 10." Section 5 required the company "to provide the Minister of Indian and Northern Affairs with evidence of financial responsibility as required under subsection 8 (1) of the Arctic Waters Pollution Prevention Act in the amount of $10 million in the form of a demand note(s) for each well." The environmental protection requirements included standard clauses (similar to those for the deepwater operations in the Beaufort Sea and ice island operations in the Sverdrup Basin) in compliance with the requirements of the various acts and a brief description of the topics to be considered in environmental studies.[3]

Although the approval in principle was granted in anticipation of drilling in 1975, Norlands did not make application for a drilling authority, nor did it explain why it did not. *Oilweek* resurrected its "political-pressure-by-environmentalists" hypothesis as one of the possible reasons for the delay:

While operators privately mention that the more rabid environmentalists in the North have either been stilled or have grown more realistic, there are isolated pockets where this is not true, leaving the northern offshore issue especially something to be handled gingerly. This is quite possibly one of the reasons, along with the ones causing money raising difficulties for Panarctic that no word is emanating from Norlands Petroleums Ltd., the operator of Magnorth's holdings of offshore Arctic acreage. Norlands/Magnorth received approval in principle for a drillship operation in the Lancaster Sound area last year but plans, if they exist, are not being talked about.[4]

Whatever the cause of the delay in the application for a drilling authority, the Norlands' proposal presents a textbook case of the conflicts between resource development, environmental protection, and the use of renewable resources by native peoples. In Arctic terms, the Sound and the adjacent areas are teeming with life, particularly marine mammals, polar bears, waterfowl, and sea birds. So ecologically rich is the Sound that the Canadian Committee of the International Biological Program has proposed that the Sound be established as a major ecological preserve.

The Environment of Lancaster Sound and the Northwest Passage

Lancaster Sound lies at the eastern end of the Northwest Passage, between the north end of Baffin Island and the south side of Devon Island. With M'Clure Strait and Viscount Melville Sound, it forms the northern part of the Arctic lowlands geologic province, one of four major geologic provinces occurring in the Arctic Islands sedimentary area. Although little information on the oil and gas potential of the region is available, the activities of petroleum firms in the area suggest that exploration prospects are attractive. Historically, the Sound is best known as the entrance to the Northwest Passage. The eastern approach was mapped by William Baffin in 1616, but it was not explored until 1819 when Captain Parry entered the Sound and sailed through the Passage as far as M'Clure Strait, the passage which separates Banks and Melville Islands.[5]

Geographically, the southern perimeter of Lancaster Sound is defined by the north shore of Bylot Island and Baffin Island, including the northeast corner of Somerset Island, while the southern coast of Devon Island constitutes the northern boundary marker. Principal water connections include Barrow Strait to the west, Prince Regent Inlet, Admiralty Inlet, and Navy Board Inlet to the south, and Baffin Bay to the east.

Over northern Ellesmere Island winter temperatures are usually in the -30°F to -35°F range. However, the moderating influences of open water from Lancaster Sound and Baffin Bay raise temperatures along Canada's eastern Arctic margins by 10°F to 15°F throughout this season.[6]

Lancaster Sound has a longer ice-free season than other parts of the Northwest Passage. M'Clure Strait at the western end of the Passage for example, seldom has more than 60 ice-free days and is periodically blocked by ice throughout the entire year. The dynamics of Arctic ice-water relationships were studied recently using the "Quick-look ERTS imagery" from the Canadian Centre for Remote Sensing, Prince Albert Satellite Station.[7] The coverage was for the entire Canadian Archipelago and adjacent Arctic Basin. From the report it is possible to piece together a one-season picture of ice and water interaction in the Northwest Passage:[8]

In northern Baffin Bay breakup began with a major lead opening 100 km (66 mi) off the Baffin Island coast and the enlarging of North Water westwards into Lady Ann Strait and northward towards Smith Sound.

The westward progression of breakup into Lancaster Sound was preceded by open leads along its northern and southern margins. Breakup reached 85W in mid May, 90 W in mid June, and 100 W (Barrow Strait) by the end of July. The Sound did not clear until September when the advection of ice from the channels to the north and south ceased.

The westward progression of breakup in Lancaster Sound permitted ice floes to move eastward throughout the earlier part of the season. Toward the end of June ice floes moving into Prince Regent Inlet from Lancaster Sound were converging on floes trying to leave the Inlet after breaking away from the edge of the fast-ice. A similar behavior was observed in mid-July at the entrance to Admiralty Inlet. During the first part of August the movement of ice in both inlets was apparently random with no consistent direction to the ice drift.

The eastward drift in Lancaster Sound varied between 9.5 and 55 km/day in August. Earlier in the season the floes were moving at 11.5 — 22.5 km/day with the speed increasing toward the south. As the season progressed, the eastern ice drift on the southern side of the Sound increased in speed while on the northern side of the sound the drift became westward.

Breakup westwards from Lancaster Sound reached Viscount Melville Sound in early August, although minimum ice conditions were not seen until mid to late September (2-7/10 cover off northern Victoria Island and McClintock Channel and ice free conditions south of Melville Island). M'Clure Strait broke up by the end of July but similarly was only partially clear until mid-September. However, the fast ice in the northern end of the strait persisted until early September.

The ice of the southern straits, the historical "Northwest Passage" thinned in situ throughout June. Breakup occurred mostly from the west in early June and the passage cleared by the third week of August. However a medium ice cover persisted in McClintock Channel throughout the season.

Minimum ice extent in many areas occurred during early September; Baffin Bay, Lancaster Sound . . . and the historical "Northwest Passage" were clear of ice. By 28 September ice had once more drifted into Barrow Strait. By 2 October Jones Sound, Wellington Channel and M'Clure Strait also had heavy ice cover.

Imagery became infrequent in October and ceased, even in the southern portion of the study area by 9 November due to encroaching winter darkness. At this time ice conditions had returned to Prince of Wales Strait, McClintock Channel, Franklin Strait and the Gulf of Boothia.

Lancaster Sound is a major staging centre, travelway, and breeding area for several species of aquatic mammals. Along the east coast of Baffin Island, between Navy Board Inlet and Cape Adair, there are an estimated 66,300 ringed seals. A number of areas from Cape Hay on Bylot Island in the east to Prince Leopold and Somerset islands on the west are breeding areas for ringed seals. Other seal species, including the bearded, hooded, and harp seals occur in lesser numbers. The Sound is an important summer migration route for the harp seal.[9] This seal returns to the Atlantic Ocean when winter comes to the Arctic and is killed by sealers in March off the northeast coast of Newfoundland and in the Gulf of St. Lawrence.

Bylot Island at the southeast end of the Sound (already designated as a migratory bird and waterfowl sanctuary) is one of the best studied areas within the region. Its diverse fauna offers an excellent example of the wildlife resources that inhabit the region.[10] The western side of the island contains a critical nesting area for nearly one-third of the world population of the greater snow goose. Huge colonies of cliff-nesting sea birds such as murres, kittiwakes, and guillemots are located on several of the capes and cliffs. One of the few detailed population appraisals, conducted in 1957, revealed that approximately 400,000 pairs of thick-billed murres were nesting in only one of several colonies at Cape Hay.[11]

Other important sea bird colonies exist on Prince Leopold Island and on the northeast part of Somerset Island. In 1961 it was reported that more than 350,000 thick-billed murres, 100,000 to 180,000 kittiwakes, about 150,000 fulimars, about 4000 black guillemots, and about 2000 glaucous gulls were nesting in the area.[12] The large murre colonies of the High Arctic make it the most abundant sea bird in the Northern Hemisphere. Murres play an important role in fertilizing the Labrador current which flows south, eventually providing nutrients to the prolific fishery of the Grand Banks.[13]

Walrus are not nearly as plentiful as seals but they also utilize Lancaster Sound as a summer migration route. The Wollaston Island area off of Bylot Island at the entrance to Navy Board Inlet and between Cape Clarence (Somerset Island) and Cape Charles Yorke (Bordeur Peninsula) or the entrance to Prince Regent Inlet are major staging areas.[14] The walrus stocks of Baffin Bay, Lancaster Sound, and Jones Sound are important since they serve as natural reservoirs for the species.[15]

Polar bears are also abundant throughout the Lancaster Sound area. It appears that it may be the route for animals which migrate to or from Baffin Bay. Three-quarters of the southern coast of Devon Island is listed as a critical habitat throughout the year. Denning sites have been confirmed at Cape Coutts and other locations.[16]

Summer is also the whaling season. The population of narwhal was estimated at 5000 in 1970 and the population of white whales (beluga) at 2000 for the Mackenzie Delta area and 10,000 for western Hudson Bay. By June, migrating narwhal from Baffin Bay as well as white whales mass along the floe-edge between Cape Crawford and Cape Charles Yorke at the mouth of Admiralty Bay. In July and August, they follow the break-up of ice along the west side of Admiralty Inlet.[17]

The Inuit communities of Arctic Bay, Pond Inlet, and Resolute make extensive use of the animal resources of Lancaster Sound and other related marine areas which together make up the eastern approach of the Northwest Passage.[18]

The Approach to Petroleum Exploration in the Northwest Passage

The most active land holder in the Northwest Passage is Magnorth Petroleum Ltd., with a total of 14.4 million acres in exploratory permits in the Arctic Islands, most of it in the Northwest Passage. In an arrangement that is common among petroleum exploration firms, Magnorth will give 25% of its exploration interests to Norlands in return for Norlands' exploration work on the Magnorth permit areas (this type of arrangement benefits Magnorth by allowing the company to meet the work obligations on its vast permit areas). The hidden link is Northern Natural Gas Co. (Nebraska), an American gas distributing company which owns Norlands Petroleum and operates the firm as its Canadian exploration arm.[19]

In 1973, Magnorth and Norlands drew up a five-year plan for exploration in the Northwest Passage. The programme was originally estimated to cost $10 million and, according to *Oilweek*, the original arrangement included plans for test drilling "possibly as early as 1976,"[20] By the end of the 1973 season, Magnorth-Norlands had run seismic surveys over more than 13,000 miles of the Northwest Passage, including intensive coverage of Lancaster Sound. It appears that the 1973 seismic work pointed to some very promising geological structures in the Sound. In spite of the extraordinary water depth, the short drilling season, and difficult ice conditions, Norlands moved its drilling schedule ahead by at least a full year, proposing to DINA that drilling begin in 1975.

There is little public information on Norlands' proposal to drill in Lancaster Sound. The total public information released by DINA consists of one letter conveying the approval in principle and an "Attachment" stating the terms of approval. Private contacts confirmed that Westbourne Engineering Co. of Calgary had conducted the engineering feasibility study and that F.F. Slaney Co. of Vancouver had made an initial assessment of the potential environmental effects associated with the drilling system.

Since Norlands' application was submitted in February 1974, the environmental assessment was probably researched and written in the latter half of 1974. If so, it would have been going on at approximately the same time that Slaney prepared the assessment of Offshore Hecla for Panarctic.[21] Given the inadequacies of that assessment, can one reasonably expect that the Lancaster Sound study was any more thorough?

Following receipt of Norlands' application for approval in principle, DINA requested DOE to review the application. DOE recommended that approval be deferred. In fact, the matter was of such concern within the Department that the Minister was apprised of it by his Senior Assistant Deputy Minister, whose memorandum anticipated that DINA would react unfavourably to the recommendation, as indeed DINA appears to have done. The memorandum tells a great deal about the way decisions on petroleum exploration in the Arctic are made and about the uneasy relations between DOE and DINA:[22]

Memorandum to the Minister

Subject: Environmental Responsibilities North of 60 — Offshore Drilling in Lancaster Sound

I am concerned that Environment Canada advice to the Department of Indian and Northern Affairs may be ignored. If this happens, and an environmental disaster subsequently occurs, DOE would be subject to strong criticism.

As you are aware, Fisheries and Marine Service has been deeply involved, along with the other DOE Services, in carrying out the Departmental responsibility for protection of aquatic ecosystems in respect of offshore drilling North of 60.

As part of their normal procedure, DINA officials, upon receipt of applications for Approval in Principle to drill, request DOE and other federal departments to provide consultations. For each such request, an indepth technical review is conducted with input from all DOE Services and a response to DINA is prepared, noting environmental concerns and advising on constraints necessary to meet these concerns.

We have recently been requested to provide consultation on an application from Norlands Petroleum Limited for Approval in Principle to drill during 1975 in Lancaster Sound (Eastern Arctic). The proponent wishes to operate a Havdrill class drillship (similar to the Pelican in use off Labrador) in these ice infested waters. The site proposed is in 3000 feet of water, a depth in which no well, worldwide, has yet been drilled.

While much baseline data on the aquatic ecosystem of Lancaster Sound is lacking, we can state that the environmental consequences of an oil spill to the massive bird populations (Canadian Wildlife Service tells us that 50-60% of all seabirds in the Eastern Canadian Arctic are in Lancaster Sound), fish and marine mammals of the area would be beyond compensation.

Because of the water depth proposed and the lack of an adequate real-time environmental hazard prediction system for weather, icebergs and ice floes, DOE considers the risk of an oil spill from this operation to be significantly increased above more conventional operations.

Were a release of oil to occur as the result of a blowout during the latter part of the drilling season, the proposed backup system (off Labrador) would not be able to drill a relief well and contain the blowout until the following drilling season. In addition, there is at present no oil spill clean up technology, either within government (MOT) or industry, which would be adequate for application in Arctic waters.

For these reasons, it was recommended to DINA that Approval in Principle be deferred until conditions which would address our environmental concerns were met. The completion of the Beaufort Sea Studies in 1976 may enable some of these conditions to be met (e.g. the Real-Time Environmental Hazard Prediction System being developed by the Atmospheric Environment Service).

I understand that DINA wishes to make a decision imminently. We are concerned that our advice may be ignored. If this happens, and an oil spill subsequently occurs with what would certainly be disastrous environmental consequences, DOE would be subject to strong criticism.

It is of interest that the description of the hazards and risks of offshore drilling in this area by a DOE senior official and professional engineer are at least as strong as those used in an article in *Nature Canada,*[23] which was subsequently described by *Oilweek* as "paranoid".[24] At any rate, it is obvious that Lancaster Sound is one of the most important ecosystems in the Arctic. It is as important to the eastern Arctic as the Beaufort Sea is to the western Arctic. No drilling should be undertaken there before the potential environmental and social effects are thoroughly assessed and publicly reviewed. The non-development of the Sound is an alternative which warrants serious consideration by both white and Inuit society.

References and Notes

1 *Offshore, Shell's 2150 ft. well off West Gabon is the world's deepest,* 20 June 1974.

2 Letter to J.T. Raleigh, President, Norlands Petroleum from A.D. Hunt, Assistant Deputy Minister, Department of Indian and Northern Affairs, 8 August 1974.

3 Attachment I (to letter, n. 2), *Approval in principle to drill wildcat wells in Lancaster Sound.*

4 *Oilweek, Mackenzie Delta hums with activity despite exploration and political setbacks,* 3 March 1975.

5 Taylor, A., *Our polar islands — The Queen Elizabeths, Canadian Geographical Journal,* June 1956.

6 Canada, Dominion Bureau of Statistics, *The climate of the Canadian Arctic,* augmented reprint from *Canada Year Book* (Ottawa: Queen's Printer, 1967).

7 Ramseier, R.O. et al., *Ice dynamics in the Canadian Archipelago and adjacent Arctic Basin as determined by ERTS-1 observations,* Proceedings: International Symposium on Canada's Continental Margins and Offshore Petroleum Exploration, Calgary, 29 September — 2 October 1974.

8 Some licence was taken in drawing together the information on ice dynamics (see n. 6) which relates most directly to Lancaster Sound and the Northwest Passage. In some cases the order and structure of paragraphs was changed.

9 Bissett, D., *Northern Baffin Island: an area economic survey,* (Ottawa: DINA, 1967), vol. 2.

10 Canadian Wildlife Service, *Critical wildlife areas, Lancaster Sound, 2036.* Arctic Ecology Map Series and related text (Ottawa: Department of the Environment, 1972).

11 Tuck, L.M., *The Murres — their distribution, populations and biology — a study of the genus Uria* (Ottawa: Canadian Wildlife Service, Canadian Wildlife Series No. 1, 1960).

12 Barry, T.W., *Sea-bird colonies of Prince Leopold Island and vicinity, Canadian Field Naturalist,* vol. 75, 1961.

13 Tuck, see n. 11.

14 Bissett, see n. 9.

15 Mansfield, A.W., *The walrus in the Canadian Arctic* (Ottawa: Fisheries Research Board of Canada, 1900), Circular No. 2.

16 Canadian Wildlife Service. See n. 10.

17 Canada, Fisheries Research Board of Canada, *F.R.B. Review 1973* (Ottawa: Department of the Environment, 1974).

18 A study of land use and occupancy was conducted recently to document the use of land and resources by the Inuit people of the Canadian Arctic.

19 *Oilweek, Historic Northwest Passage promised North Sea potential,* 8 October 1973.

20 *Oilweek,* see n. 19.

21 The preliminary environmental assessment prepared by F.F. Slaney and Co. for Offshore Hecla is discussed in Chapter 5 on drilling from ice islands.

22 The copy of the memorandum we obtained had been cropped in copying and had no date on it.

23 Pimlott, D.H., *The hazardous search for oil and gas in Arctic waters, Nature Canada,* October/December 1974.

24 *Oilweek, Pace of Arctic islands exploration hinges on discoveries and regulations,* 3 March 1975.

Chapter 8

The Hazards and Risks of Offshore Drilling

The hazards of exploring for and producing oil and gas in the offshore waters of the Arctic are awe-inspiring. The risks to the environment of a major oil spill from a blowout, or an accident of another nature, are frightening. Drilling operations face a formidable array of environmental forces: ice in many different forms; a short ice-free season; intense, prolonged cold during the winter; powerful storms which are difficult to predict; and in some areas, very high pressures in geological formations. Nevertheless, almost the entire Arctic offshore area is blanketed by oil and gas exploration permits.

The outcome of exploration and development activities is critical to the Arctic's natural ecosystems and their use by Inuit. In many Arctic regions, an accidental blowout could spew oil or gas for almost a year before a relief well could be drilled. The consequences of a major oil spill under Arctic circumstances would be sweeping, and the destruction of sea birds, marine mammals, and other aquatic organisms could be beyond recovery in areas such as the Beaufort Sea, Hudson Bay, and Lancaster Sound.[1] Despite these hazards, it is difficult to believe that massive oil spills will not occur in Arctic waters during the next decade.[2]

The Draft Memorandum to Cabinet on offshore drilling (Appendix 1) provided virtually no information on environmental risks, but said that these were considered to be few. The memorandum said, in reviewing offshore drilling throughout the world, that "Off-shore oil and gas development precedents in the ice-infested waters of Cook Inlet, Alaska, the cold northern waters of the North Sea and the cold southern waters of the Bass Strait of Australia illustrate that discoveries can be developed without apparent damage to related marine ecosystems."[3]

It is misleading to use the North Sea and the Bass Strait of Australia as precedents. The problems to be encountered in the Beaufort Sea are primarily ice problems, and so only Cook Inlet is a valid anology. But, in fact, ice conditions in Cook Inlet are relatively mild. R.D. Johnstone of Atlantic Richfield described Cook Inlet's ice and weather conditions, at the Northern Canada Offshore Drilling Meeting:

The weather conditions in the Cook Inlet are not considered severe when compared to our Arctic offshore area. There is an ice period from December through April when the average temperature is 20°F. There will be cold periods lasting two to three weeks when the temperature will reach 20°F to 30° below zero. The ice cover reaches a thickness of two and a half to three feet and extends to a point below Kalgin Island. The ice moves back and forth with the change in tides reaching a speed of four to five knots. This motion keeps the ice broken up but some ice pans will be ten to twenty acres in area.

The ice leaves the Inlet in April and the Inlet remains ice-free until the freeze-up in November. During this period there is fairly good weather with some rain and spring fog.[4]

Oil in Arctic Marine Ecosystems[6]

The oil spills which resulted from the grounding of the *Torrey Canyon* and the

Arrow, and from the blowout at Santa Barbara, alerted the world to the potential consequences of large crude-oil spills. But these spills occurred in benign environments where the problems of cleaning up oil were relatively simple ones. Under Arctic conditions, there simply is no existing technology and no equipment available which can be considered at all adequate for cleaning up a massive oil spill.[5]

Crude oil is a complex substance which varies a great deal in composition from one oil reservoir to another. The toxicity of a particular crude is a function of its degree of solubility and of the aromatic hydrocarbons which it contains. As soon as crude oil gets into the sea it begins to be modified by weathering processes: evaporation, dissolution, microbial action, oxidation, and photo-chemical reactions. The rate of weathering is dramatically influenced by several physical factors such as temperature, light, winds, tides, currents, waves, and the presence of ice. Temperature is of primary importance to biodegradation processes, but the presence of pack ice could be very important in reducing the evaporation of the volatile compounds which are quite toxic. The toxicity of the residue changes considerably during different stages of weathering.

Research to determine how these processes work in the Arctic has just started, so results obtained by the Beaufort Sea Project will undoubtedly be subject to many qualifications. For example, preliminary results indicate that oil trapped in pockets under the ice does not weather significantly and months later may be just as toxic as freshly spilled oil when it finally melts out. On the other hand, it appears that when it finally surfaces bacterial action may be effective in degrading it, even at temperatures in the 0° to 5°C range. This is encouraging because only two years ago it was postulated that oil would remain in Arctic seas for a long time because low temperatures reduced or stopped microbial processes.[7] Under open-water conditions the low-boiling (volatile) components evaporate and dissolve within two days. Those with higher boiling points are degraded by microbes and by chemical oxidation and take much longer, sometimes years, to disappear from an ecosystem. The heavy residues either end up in sediments or form the tar lumps which occur widely at sea and are found on beaches all over the world. Oil compounds which end up in sediments may retain their toxic characteristics for a long time and remain there indefinitely or, if they are in areas of shallow water which are subject to churning by either wind or current, they may recontaminate an area long after the spill occurred. Sedimentation processes involving crude oil or residual compounds in coastal areas could have a direct impact on fish spawning grounds or nursery areas.

Oil and gas rising from a blowout on the bottom of the sea under high pressure rise in a stream of fine droplets entrained in a circulating plume. Some of this oil will be dispersed through the water column in a semi-stable emulsified form.[8] Some of the oil which reaches the surface can also end up in the water column, as a result of mixing during storms. Virtually nothing is known about what will happen in the Arctic to oil in this form, or what effect it will have on the ecosystem over the long term. It is one of the important areas for research in the immediate future.

Effects of Oil on Individual Organisms and Populations

Public understanding of the effect of oil spills on animals is based largely on the great number of birds killed by some spills.[9] But the impact of oil spills can be much more important than the deaths of birds or mammals. Spills can also have long-term sublethal effects which, by influencing the physiology and behaviour of organisms, can be even more important than deaths caused directly by oil. Habitat changes can be subtle and unnoticed but may reduce both the diversity and abundance of animal and plant communities.

In temperate regions there is always great concern about oil fouling shorelines, partly because human economic and recreational interests are often involved and partly because there are diverse biological communities in the intertidal zone. The latter rarely applies to the Arctic because "in most areas of the Arctic the intertidal zone is depressingly barren and damage to invertebrate fauna will be minimal."[10] The Arctic intertidal zone's barrenness is sometimes attributed to ice-scouring and to the extreme winter temperatures. The most important question in a consideration of the potential effect of oil spills on fauna is not simply, what animals will be killed? but rather, what animals will be killed, and what will it do to populations of the various species over the long term? I will attempt to review the kinds of problems which could result from a major oil spill and, in some instances, the effect they might have on the species involved.

Birds and Mammals

In Arctic waters, polar bears, marine mammals, waterfowl, sea birds, loons, and shorebirds are the animals which could be most seriously affected by minor or major oil spills. The behaviour of most of these animals endangers them when they come in contact with intensive human activities. They are all very mobile, and many species concentrate in large numbers in particular areas at predictable times; they are highly vulnerable to oil spilled at or carried by currents to these areas. There have been cases where migrating birds apparently have been attracted by pools of oil on the surface of the water or ice. In 1957, for example, 450 migrating ducks and geese landed in a pool of crude oil spilled by Imperial Oil on the Mackenzie River near Norman Wells. All the birds were killed.[11]

As oil works its way to the surface, it will concentrate in the open leads which form in response to pressure and to seasonal changes. It appears that Arctic migration patterns of many species have formed around the chain of leads which occur along the Beaufort Sea coast. For waterfowl and sea birds in migration patches of open water are of critical importance for both feeding and resting.

The waterfowl most common in the Arctic are whistling swan, white-fronted goose, snow goose, brant, and several species of diving ducks. These include scaup, old squaw, Pacific, and king eiders, white-winged and surf scoter. Estimated populations of swans, geese, and brant which use the Arctic coastal migration route are about 400,000. The number estimated for eiders is one million and for scaup two million. There are not even good guesses at the present time for the other species.

Snow geese would probably be the least vulnerable of all waterfowl to oil spills because they feed a great deal on land and do not use the salt water nearly as much as do the diving ducks. If conditions were favourable at the time of a spill, flocks could conceivably pass through the damaged area without being seriously affected. In the Beaufort Sea, storm tides would seem to pose their greatest threat, because when storms occur, many of the areas important to geese are flooded by the sea. But if kills did occur, they could be very extensive because of the frequent sizeable concentrations of birds along the coast. For example, the importance of the mouth of the Mackenzie River to migratory waterfowl was illustrated by the F.F. Slaney study for Imperial Oil in 1972. During the spring migration, peak numbers were reported on 3 June and nine concentrations of waterfowl were observed: "the largest of these was at the mouth of the East Channel of the Mackenzie River where over 10,000 waterfowl . . . were associated with areas of exposed vegetation and open water on low-lying, sedge-covered islands." Most of these birds were snow geese. During the fall migration the area was used mostly during the first two weeks of September: "A group of approximately 6000 birds was observed on 4 September . . . on 7 September 11,000 birds were located . . . . On 13 and 16 September, concentrations along the coast (totalled) 27,000 and 23,000 respectively," again mostly snow geese.[12]

Waterfowl ecologist T.W. Barry referred to the importance of the Babbage River Delta, Blow River Delta, and Shallow Bay of the Mackenzie River Delta to waterfowl: "During the autumn these places are staging areas for approximately 75,000 snow geese, 25,000 brants and 10,000 white-fronted geese. The three delta sites are extremely important nesting grounds for whistling swans, the densest swan nesting I know of is in the flats from Blow River to the West Channel of the Mackenzie on Shallow Bay."[13] These concentration areas are all in the part of the Beaufort Sea where intensive exploration for oil and gas is being, or will be, conducted from artificial islands. A blowout could have serious consequences for North American populations of snow geese and swans.

Brant (Atlantic and Pacific) use offshore areas extensively for feeding and resting. As many as one-third or more of the total population of Pacific brant apparently collects, at times, in the southern Beaufort Sea. Concentrations are greatest during bad weather when the birds stop over to wait for better conditions to continue their migration. In the eastern Arctic, Foxe Basin and parts of Hudson Bay are important both in migration and nesting.

Brant would be endangered by oil spills during the nesting season and during migration because they usually nest very close to the high-tide line on flat areas or low-lying islands. It would appear that oil spills in Arctic coastal areas would constitute a serious threat to populations of both species. Eider ducks, scaups, old squaws, and scoters spend a great deal of their time at sea and could also be hit hard by oil spills in the Arctic. Barry made this statement about the Tuktoyaktuk area:

Currents and tide-rips along the coast are strong and changeable. In general, they parallel the shore in a north-east to south-west direction, and even on a calm day an oil spill would spread rapidly. The shallow sandy bottom, which in some places extends 25 km or more from the shore, the lagoons behind the barrier beaches and the drowned lakes of Tuktoyaktuk Peninsula are important for huge numbers of sea birds found there during open water. King and common eiders pass this part of the Beaufort Sea in almost continuous migration, beginning as early as late April when the first male eiders appear in open leads and lasting until well into October. They would certainly suffer from oil spills. During July and August the waters and bays are important moulting grounds for old squaw ducks, surf and white-winged scoters and scaup ducks. I can only guess that their numbers might be 600,000 or more.[14]

The murre is the most common sea bird of the Arctic.[15] At least a million birds nest around the Polar Basin and there are several million more in the eastern Arctic. In the late 1950s, a colony on Digges Island at the northeastern tip of Hudson Bay was estimated to contain between two and three million birds; another at Aktopok Island had close to a million nesting birds. In the Arctic, murres nest on steep cliffs on the edge of the sea and feed within a radius of five to ten miles from the colony. The young make a short flight from the cliffs to the sea before they are capable of normal flight. These colonies are extremely vulnerable to oil spills. A spill in the vicinity of the Digges Island colony, for example, could eliminate the entire colony in matter of days.

There is only one murre colony in the western Arctic. Barry said of it that "we have been anxious for the safety of the only murre colony in the western Arctic. The fuel storage tanks for the main Dewline site at Cape Parry are only a few hundred yards from the bird cliffs where less than one hundred pairs of murres nest. A spill, even of diesel oil during resupply, could wipe them out. There are no murre colonies for more than 1600 km east or west from Cape Parry."[16]

Gulls, jaegers, terns, shorebirds, and loons are among the groups of birds whose population could conceivably be affected by oil spills. Range maps in *The Birds of Canada*[17] and the *Arctic Birds of Canada*[17] show that more than 20 species of shorebirds nest in the Arctic. Many of them use marine beaches, or low-lying flats close to the sea, during migration. All four species of loons (Arctic, common, red-throated, and yellow-billed) are sometimes found in the Beaufort Sea during migration. Flocks of 500 or more are not uncommon. Some loons also make feeding forays to coastal bays and inlets during the nesting season.

Very little research has been done on any of these birds, and our ignorance of what may constitute important or critical habitat during their migration is profound. In such cases, assurances that adequate assessments of the potential environmental impact of oil spills can be made in one or two field seasons sound particularly empty to ornithologists and naturalists.

When oil spills occur, the fate of birds is always brought to the public's attention. Birds frequently concentrate close to shore, they are very pitiful and obviously doomed creatures when they are covered with oil, and after death they usually float on the surface of the water and are relatively easy to locate and count. It seems likely that losses of marine mammals and polar bears would be much more difficult to document. These animals are more widely dispersed and their bodies frequently do not float after death. There is only limited evidence of marine mammals killed by oil, for example, juvenile harp seals which were thought to have encountered an oil slick in the Gulf of St. Lawrence. It was believed that at least 3000 to 5000 seals were involved and many died: "Dead oiled seals were found on ice floes, along the beaches of the Strait of Belle Isle and some were observed dead in the water."[18]

An attempt was made to study the effect of the oil spilled after the *Arrow* went aground in Chedabucto Bay in 1970. The population of grey seals in the vicinity of Chedabucto Bay was much lower than usual (500 where there were usually thousands) but only 13 were found dead. On Sable Island, 100 miles away, most grey and harbour seals were either lightly or heavily oiled but less than 5% were found dead.[19] No evidence was presented in either case to prove that the dead seals had died from the effects of oil.

The observations from Chedabucto Bay and other limited evidence suggest that marine mammals are more tolerant of oil spills than birds, and many on occasion deliberately avoid oil spill areas.[20] The tolerance of seals, for at least limited periods, was indicated by a field study where six ringed seals suffered no permanent damage after living in oil-coated water for 24 hours, but in a laboratory study, three seals subjected to oil in a tank all died within 70 minutes of contact with the oil.

The lives of polar bears and seals are very closely linked throughout the Arctic. In all areas except Hudson Bay, the bulk of the bears' diet throughout the year is ringed and bearded seals, particularly the former, the most abundant marine mammal in the Arctic. Ringed seals are normally solitary animals and spotting even two together is a rare event. Occasionally they congregate in the late fall, possibly in response to concentrations of the small fish on which they feed. Ringed seals are widely distributed throughout the Arctic with higher populations in the east than in the west.[21] Possibly the wide distribution of seals would also result in a reduced impact of oil spills on populations. However, there are many unknowns. For example, they might be most vulnerable when young seals are in their birth lairs on shorefast ice: it is then that ringed seals occur in greatest densities, and if oil from a spill circulated under the ice in the Amundsen Gulf in the early spring it could conceivably result in many deaths.[22]

Polar bears' behaviour could reduce the direct impact of oil spills on them. In Hudson Bay they spend the summer on land; in the Beaufort Sea they remain with the pack ice and consequently away from the principal area where offshore drilling will occur. However, since bears are so dependent on seals for food they would be indirectly affected by reduction in seal numbers. But so little is known about currents in the Arctic, the action and behaviour of oil on, around, and under the ice, and the reaction of the animals to oil that speculation about possible effects of oil spills on marine mammals is almost pointless, and a crash research programme such as the Beaufort Sea Project simply cannot supply the missing information.

Fish and Invertebrates

Short-term laboratory experiments have provided considerable information about the toxicity of oil to fish:

Such short-term toxicity tests have been the favourite research tool of pollution biologists interested in gaining an insight into the effect of various pollutants on aquatic organisms. Useful as such studies have been in outlining the nature and general scope of pollution damage they can never be considered as more than crude first approximations. Of more importance in the long run are those subtle, noxious effects that, while they don't kill individuals outright, may nevertheless result in the extermination of the population in the course of one or several generations.[23]

A report by the Council on Environmental Quality listed five main ways that local fish populations can be damaged by oil:

(1) Eggs and larvae die in spawning and nursery areas from coating and from exposure to concentrations of hydrocarbons in excess of 0.1 parts per million. (2) Adults die or fail to reach the spawning grounds if the spill occurs in a critical, narrow or shallow waterway. Anadromous fish are particularly vulnerable to this situation. (3) A local breeding population is lost due to contaminated spawning grounds or nursery areas. (4) Fecundity and spawning behaviour are changed. (5) Local food species of adults, juveniles, fry or larvae are affected.[24]

The fish studies conducted for the Beaufort Sea Project were baseline studies to determine the occurrence, distribution, and food habits of species in southern coastal areas of the Beaufort Sea.[25] They represented the first systematic studies in the area.[26] One of the reports identified sediment sinks, lagoons, and bays and the coastal margins as areas of particular importance to fish of the region. The sinks result when opposing currents meet and drop their sediment loads as the currents slow down. The studies discussed two such areas off the Yukon coast and stated that they contain large numbers of fish and are extensively used as feeding and rearing areas.[27] In these areas oil mixed with sediments could cause problems during periods of high wind. The studies also discussed specific ways in that oil might influence fish in other critical habitat areas, and rejected the idea of using lagoons as natural areas to impound spilled oil because of their importance as spawning and rearing areas for fish.[28]

A study of oil spill effects on a few species of marine invertebrates was also conducted as part of the Beaufort Sea Project.[29] Apparently it represented the first test on Arctic invertebrates since the bibliography of the paper contained no references to research of a similar nature. On the basis of standard 96-hour lethal toxicity tests, with a variety of crude oils in seawater emulsions, it was found that the species tested were relatively resistant to crude oil except when the concentrations were very high. The tests approximated summer temperature and open-water conditions. The behavioural responses to oil were also tested for a number of species. One isopod showed no response while two others avoided masses of oil and tainted food. These responses diminished with weathered oil or if the animals were pre-exposed to oil emulsions before the experiments were conducted. In summing up his work, the author stated that the evidence suggested that massive rapid kills will be limited primarily to areas in the immediate vicinity of a blowout. However, it was pointed out that over the long-term, low levels of toxic oils, because of their sublethal effects, could have significant ecological consequences.

Physical Hazards to Offshore Operations

Physical environmental factors pose hazards to many aspects of Arctic offshore operations. They could result in oil spills at any phase of the development of oil fields. Blowouts could be significant in the exploration and production phases, but chronic equivalent losses might occur during storage and transportation operations.[30]

Where long-term climatic records are available climatologists can provide a statistical assessment of the periodicity of many phenomenon such as maximum sustained wind, maximum wave height, icing probability, and the frequency of storms of varying magnitudes. The U.S. Council on Environmental Quality has had such data interpreted in terms of equipment and structures, and can predict whether or not major oil spills are likely to be caused by natural phenomena during the life of oil fields in particular areas.[31] However, data which lend themselves to such detailed interpretation are rarely available for Arctic offshore areas.[32] In addition, reports on studies which have been conducted, either theoretical or practical, are usually difficult to obtain either because of government secrecy or of industry's proprietary interests.

In the exploration phase the principal danger is that wind, ice, earthquakes, or subsurface geological pressure will cause blowouts. Drillships or semi-submersibles could be forced to leave their drilling stations, perhaps fracturing the marine riser, and if the drill has reached an oil-bearing formation a blowout could occur. In theory at least, the blowout preventor (BOP) acts as a first line of defence against such an occurrence, but many oil spills have been associated with accidents resulting from storms.[33]

Variable geological formation pressures pose a much greater threat during drilling and cause the majority of blowouts. However, much of the data on this phenomenon are closely guarded by the oil industry. Even the two comprehensive reports on offshore drilling on U.S. outer continental shelf areas do not provide any specific information on the hazards to drilling operations posed by abnormally high formation pressures.

Ice Hazards

Floating ice is the most characteristic feature of the Arctic marine realm, and it dominates all Arctic marine and offshore activities. Either we run away from it, schedule our activities to avoid it or we completely adapt and reorganize our operations to cope with it, in every case the ice is ultimately in control. If it were not for the presence of floating ice and its peculiar qualities, industrial operations in the Arctic offshore areas would be merely an extension of offshore operations anywhere else, with the added but calculable factors of distance, isolation, and severe climate. But floating ice is an adversary of a different breed than we are used to. Where modern man has met it in his daily life or in his commerce until now, ice in or on the sea has been light enough that he could use brute-force and smash it — the traditional icebreaker approach — or transient enough that if he had patience it would go away and operations could return to "normal". But for much of the area of promising oil potential in the Canadian Arctic offshore, the normal situation is ice on the sea all the time, in such masses that the brute-force approach to dealing with it is bound to be economically unsound if not technically impossible.[34]

The major technical problems confronting offshore drilling operations in the North are associated in one way or another with sea ice, i.e. the kinds of ice found in Arctic tidal waters but not necessarily of sea water origin. Most certainly, other technical problems also exist, which are not normally experienced further south, such as brittle fracture of metals under extreme low temperatures, metallurgy in general, concreting in cold regions, submarine permafrost, etc. These would not be considered difficult to overcome, however, solutions to the problems posed by certain elements of sea ice is another matter . . . . [Ice] islands occur in sufficient numbers and size to represent a considerable menace to offshore drilling operations in Mackenzie Bay and the Beaufort Sea and also to pipelines and other installations on the seabed.[35]

Hazards from Sediments and Ice in the Sea Floor

Sea-floor sediments are the materials on which any bottom-mounted structures must be built, in which anchors must be imbedded, through which drills must be set and, eventually, warm products extracted. The Arctic sediments are most uncompromising and difficult materials for any of these purposes. Many of the technical difficulties, and . . . important uncertainties in the cost of exploratory and development drilling, and ultimately of production wells, collector pipelines, terminal or storage facilities, will be dependent on the completeness of our knowledge of this sea-floor layer and how to deal with it. The presence of ice on the water surface makes consideration of use of the sea-floor all the more important. Yet we have given very little attention to this uppermost layer of sediments, and our present knowledge of its properties is almost non-existent . . . .

At the present time a general picture of the sea floor shape is known as a result of numerous random track data that have been collected by ships . . . a very limited area has been systematically surveyed to modern standards. It is hard to convince many people that although there may be a quite dense coverage of soundings collected by reconnaissance methods, the area is not necessarily safe for passage of shipping. This feeling of security was shattered when during the Manhattan *voyage through the Beaufort Sea to Prudhoe Bay, the vessel came close to striking an obstacle that rose abruptly from the sea floor. Subsequently, system-*

atic surveys have revealed that there are a great many of these obstacles, that are now thought to be submarine pingos, and are a distinct danger to deep draft shipping.[36]

Climatic Hazards
The critical nature of the climate-ecology relationship and climate-development relationship cannot be overstressed. The harsh, cold Arctic climate has been a fundamental force in shaping the tundra, and the nature of the tundra can be easily upset by slight climatic changes. A knowledge of climate, particularly the critical extremes, is therefore basic to the understanding of tundra ecology and the prudent conquest of the Arctic by man.

The harsh climate poses many technical problems in construction, resource development, communications, transportation, tourism and indeed almost every activity within the Arctic. Techniques and standards developed for temperate latitudes are frequently unsuitable for Arctic extremes. Through a more complete understanding of the varied nature of climate and its biological effects, better design and planning are possible and activities may be prudently regulated. The Arctic paradox may then tend to disappear and many of the problems of production and conservation will be greatly simplified.[37]

Table 1 indicates the temperature characteristics of offshore areas where drilling will occur by summarizing temperature records for seven stations. Komakuk, Sachs Harbour, and Tuktoyaktuk border on the Beaufort Sea; Resolute, Isachsen, and Eureka describe conditions in the Arctic Islands, and Coral Harbour is situated on Southhampton Island at the northern extremity of Hudson Bay.

Some significant aspects of winds which would be of interest to offshore operations are lost in the treatment of the data.[37] They do show, however, that winds at Sachs Harbour and Cape Parry (Beaufort Sea), Coral Harbour (Hudson Bay), and Resolute (east end of the Northwest Passage) were significantly higher than those at Isaachsen and Eureka. Average annual speeds were also a minimum of two miles per hour higher than those at six of eight stations (Vancouver, Calgary, Edmonton, Regina, Winnipeg, Toronto, Ottawa, and Montreal) in the south. Maximum winds and maximum gusts also tend to be considerably higher in the eastern Arctic than in the south; those in the western Arctic do not show much difference. However, as pointed out earlier, the winds over water are often much stronger than those reported along the coast or at inland stations.

Wind chill factors are often extremely high for offshore areas in the Beaufort Sea and Arctic Islands. For example, the mean monthly temperatures at Sachs Harbour for January and February are -21.6°F and -23.7°F, the comparative temperatures for Resolute are -26.6°F and -28.3°F, and for Isaachsen -31.2°F and -33.5°F. At Sachs Harbour winds of 20 to 30 m.p.h. are fairly common during the period. A temperature of -20°F with a 20 m.p.h. wind results in a wind chill factor of -90°F.

Apart from general references, research did not turn up any studies of the frequency or the intensity of storms in the Arctic. The foreword to *The Climate of the Mackenzie Valley — Beaufort Sea* provides an explanation of why climatologists might be reluctant to attempt a discussion of these phenomena: "the Beaufort Sea analysis is based on extremely limited meteorological data obtained from ship observations. These data at best represent 30 days of continuous record. Data from contiguous coastal areas were used as guidance in the analysis. At the grid points, probability and extreme value estimates based on these data are likely

TABLE 1

TEMPERATURE RECORDS FROM VICINITY OF OFFSHORE DRILLING AREAS AND COMPARATIVE DATA FOR EDMONTON AND TORONTO

Climatic Record	Komakuk Beach	Sachs Harbour	Tuk	Resolute Bay	Coral Harbour	Isachsen	Eureka	Toronto	Edmonton
Yrs. of Data	12	15	13	23	27	22	23	29	30
Year Mean Daily	12.1[2]	7.4	12.8	2.5	11.4	-2.3	-2.8	45.5	37.1
Mean Daily									
October	16.9	11.1	19.5	5.5	18.2	-2.9	-7.2	49.7	41.9
November	0.4	-7.1	12.8	-11.6	1.3	-18.8	-23.2	38.1	24.5
December	-11.4	-16.0	-13.3	-19.9	-12.1	-26.5	-30.7	25.7	12.8
January	-11.8	-21.5	-17.0	-26.6	-22.8	-31.2	-33.9	20.6	5.5
February	-18.3	-23.7	-20.5	-28.3	-21.1	-33.5	-35.9	21.6	13.1
March	-12.5	-17.1	-12.8	-24.3	-12.1	-29.5	-34.1	30.3	22.2
April	0.4	-2.8	1.6	-9.5	2.5	-14.9	-17.6	43.5	39.2
Days with Frost	301	314	277	321	291	338	299	154	192
Extreme Maximum	81	70	86	65	77	72	67	101	95
Extreme Minimum	-54	-57	-58	-62	-61	-65	-64	-24	-55
Latitude	69 35	71 59	69 27	74 43	64 12	78 47	80 00	53 19	43 41

1 From *Temperature and precipitation 1941-1970: The North — Y.T. and N.W.T. Atmospheric Environment Service,* Environment Canada, Records for Toronto and Edmonton from AES records provided by the Inuvik Weather Office.

2 All temperatures in Fahrenheit.

to be low due to the paucity of data. The implication is that extreme events may occur far more frequently over the Beaufort Sea than the analysis of the data would indicate."[38] Unpredictable and severe weather conditions during the drilling season are potential hazards which offshore operators will have to face, at least in the Beaufort Sea. Severe storms are most likely to occur during the early fall, toward the end of the drilling season. Although the occurrence of severe storms has long been recognized in the area, only one attempt has been made to document the effects of a storm.

The potential hazards that weather conditions pose to offshore operations were dramatically illustrated in 13-16 September 1970, when a complex storm system moved in a southeasterly direction across Mackenzie Bay to the southern Beaufort Sea and the Amundsen Gulf. The effects of the storm were studied and described by personnel of the Department of Public Works, who at the time were evaluating the potential of Herschel Island for a marine terminal. Richard Brown of DPW and H.P. Wilson, officer-in-charge of Arctic Weather Central undertook an investigation "to determine the magnitude of both wind speed and storm surge, and the effect these conditions had on shoreline erosion and property damage." A further study was conducted the following spring of ice islands which had grounded in Babbage Bight during the storm.[39]

The advancing storm was first detected late on the night of 13 September by radar at the Dew-Line Station at Shingle Point, located approximately mid-way along the coast between Herschel Island and Tuktoyaktuk. The most severe effects of the storm were felt in this area. The storm, consisting of two distinct fronts, was estimated to be advancing at 60 m.p.h. Shortly after midnight on 14 September, winds up to 70 m.p.h. were recorded at the station. Although a warning was passed to two Northern Transportation Co. barges unloading in the area, one barge was grounded ashore by high winds while the other barge broke loose from its mooring and was later located stranded on a small island.

At Tuktoyaktuk, part of a road was washed out and about 20 feet of the peninsula was lost by erosion. Most of the damage in the Tuk area was caused by a storm tide of eight feet plus eight-foot waves. At Nicholson Point to the east of Tuktoyaktuk, about one-third of the local landing strip, located on a sandspit, was washed away by wave erosion. Tent Island, a small island near the mouth of the west branch of the Mackenzie River, was completely inundated by surging water resulting in the death of one man. Two small ships on the Beaufort Sea reported winds gusting up to 110 m.p.h. and waves of 25 feet.

Quite apart from the damage to coastal locations, the storm demonstrated that even during the generally ice-free months, sudden storms could drive the polar pack up against the coast, threatening drilling rigs and other offshore installations. Before the storm, the polar pack was more than 100 miles north of Herschel Island and Mackenzie Bay was completely ice-free. But within 36 hours the winds had driven a large amount of sea ice into Babbage Bight, including remnants of old multiyear floes and a number of ice islands.

To meteorologists, the September 1970 storm was a complex phenomenon, but more important than the meteorological event was the admission that for the immediate future, it might not be possible to forecast such storms. The chief of Arctic Weather Central said that "After consideration of the nature of the storm, it seems inescapable that we will continue to fail in such cases until our science and forecasting practice make sufficient advances." He continued, "This storm is a reminder of the fact that most Canadian meteorologists are woefully weak in oceanography. There can be no doubt that if we are to do a proper job of marine forecasting, we ought to have a far better knowledge of oceanography than we now possess. This is of particular concern now to the Arctic Weather Central because of the fact that we will, in the near future, be forecasting for offshore drilling, and later probably for the construction of gathering systems."[40]

Some investigators suggested that storms of similar duration and strength could be expected only once every 25 years. However, it was also suggested that the September 1970 event should not be regarded as atypical.

In one respect this storm was quite conventional in that it occurred within the stormy period of September through November. Furthermore, this type of storm likely occurs more frequently than suspected. In many cases, the trajectory of the low center is similar but displaced farther north where there is less open water and fewer reporting stations. A cursory inspection of the historical weather map series in conjunction with the available ship data would indicate a seven-to-nine year return period, although the last storm of similar severity along the coast occurred in 1944.[41]

The Arctic inversion is another northern weather condition described as "an increase in temperature with height." It is caused by the negative radiation balance over the snow and ice surfaces which is present during the greater part of the year. The character of the radiation balance is due to the extreme regime of day and night, low solar elevation, and the high albedo. Low-level atmospheric radiational cooling is frequently combined with dynamic heating aloft caused by subsidence of warm air advection, and this tends to maintain and intensify the inversion structure. The inversion persists throughout the year northward of the Arctic coast, but along the Mackenzie Valley the air mass becomes considerably more unstable during the period April through October.

Inversions play a major role in atmospheric pollution. An Arctic inversion may act as a lid or barrier to the dispersion of pollutants thus trapping them near the ground. This causes irritation and visibility reduction. Smoke plumes from chimney stacks will frequently remain at a fairly constant height under inversion conditions and drift away from the source. We may not assume this to be the case at all times, however, since it may be shown that the height of rise and amount of dispersion are governed by the exit temperature of the pollutant, the intensity of the inversion, and the wind speed.

Dangerous pollution conditions develop during calm or light winds and under pronounced inversions. The Arctic has these conditions far more frequently than southern regions. An inversion is present nearly all of the time in winter and the only way of clearing a polluted area in the Arctic is through brisk winds.[42]

Contingency plans usually include a plan to burn oil and gas escaping from blowouts. Because of inversions, implementation of such plans could result in severe air pollution problems.

Abnormal Pressure in Arctic Petroleum Basins

Zones of abnormally high geostatic pressure occur in the sedimentary rocks of part of the Mackenzie Delta, Beaufort Sea, and in some areas of Arctic Islands.[43] Gas hydrates (frozen gas and water) under permafrost and usually above 3600 feet constitute a hazard in some areas of the Mackenzie Delta/Beaufort Sea area, and are "something of a monster that northern wildcatters are just learning to control."[44] The problem of gas hydrates is unique to the Arctic, although abnormal reservoir pressure occurs in petroleum basins in a number of areas around the world.[45] *Oilweek* gave this account of gas hydrate problems:

Gas collects in hydrate form under the permafrost and generally above 3600 feet. Starting at a minimum depth of 650 feet a latticework of frozen gas molecules and water (ice) builds up at the rate of one gas molecule to six water molecules. Gas hydrates offer much more gas concentration than free gas. One cubic foot of hydrate can contain 170 cubic feet of gas. When warm drilling mud comes into contact with the frozen hydrate, gas can be released at exponentially increasing pressures. This caused delta drillers a pressure problem which had nothing to do with geopressure.

There is no inherent pressure in the crystalline hydrates, but with the introduction of any thawing possibility, gas will be released in quantities that can build pressures very quickly. In practice, high collapse strength steel casing is needed to contain a hydrate zone after a few days drilling.[46]

Geopressure zones are believed to be caused by rapid burial of impermeable shales by sediments, and by tectonic stresses which lead to the interconnection of shallow and deep strata through faults or fissures.[47] In the Beaufort Sea, pressure areas encountered in drilling from artificial islands are the result of rapid sedimentation. Impermeable shales prevent the escape of subsurface fluids, and pressure increases as more and more sediments are deposited in the Delta. The excess fluid prevents the formation from becoming completely compacted and so the fluid supports part of the pressure which is exerted by the overlying sediments.

The hydrostatic gradient — the increase in reservoir pressure with depth — varies from approximately 20 psi to 100 psi for each depth increase of 100 feet (0.2 to 1.0 psi/ft). The average, or normal, increase is generally considered to be approximately 0.46 psi/ft. It is referred to as "normal pressure for depth" and is the hydrostatic pressure exerted by a column of sea water.[48] The majority of petroleum reservoirs have pressure gradients above the hydrostatic gradient, but only a small percentage have fluid pressures which approach a geostatic gradient of 1.00 psi/ft. A fluid pressure gradient of 1.3 psi/ft. in Pakistan was referred to as being extra-high pressure.[49] Any pressure greater than the weight of an equivalent column of water from the reservoir is referred to as excess or abnormal pressure.

An integral part of drilling is the use of drilling mud to prevent blowouts by counter-balancing formation pressures and preventing oil or gas flow as the drill bit penetrates the formation. The drilling mud is pumped down the drill pipe (or string) into the hole, out through the drill bit, and back to the surface through the annular space between the drill string and drill hole or casing. It removes the cuttings from the face of the bit and carries them to the platform for disposal. Because of the considerable variation in formation pressure, the force exerted by the drilling mud on the formation must be varied by changing the composition of the mud and the pumping rate.[50]

Drilling wildcat wells into unknown geological formations can be particularly hazardous. Unexpected penetration into high-pressure zones can cause blowouts because of the difficulty of increasing the weight of the mud column rapidly enough to compensate for the increased pressure. A particularly vulnerable part of the drilling operation occurs during trips — moving the drill string in and out of the well bore — when the loss or gain of drilling mud is more difficult to monitor.[51]

Little or no information is made public by oil companies about the occurrence of abnormal pressure gradients in cases where they have been controlled and the wells completed. Limited information was made available about Immerk in December 1973 because in that case drilling had to be discontinued at 8883 feet instead of at 15,000 feet and some explanation was required. However, Imperial's statement, as quoted in newspapers, was very terse: "The well will be abandoned at the present depth because abnormally high formation pressure precluded further safe drilling." Last March, a company official provided the first public insight on what had actually led to the abandonment of the well: "Drilling stopped December '73 at 8883 feet when mud gas reached 200 units. Mud weight of 18.2 pounds was tried. It was within the one-pound safety limit to assure formation integrity but left no margin of safety for further formation pressure increases. Daw stressed Immerk was completely under control at this point but lack of further pressure leeway prompted the decision to stop drilling."[52]

The story added information which indicated that one pound of mud per gallon was equivalent to a pressure of 500 psi or a hydrostatic gradient of just over 1.0 psi/ft. If so, it indicates that the pressure at Immerk was in the high range of pressures recorded throughout the world and at, or very close to, geostatic pressure. Apparently Imperial encountered many problems while drilling Immerk as a result of the high pressure. The drill became differentially stuck (caused by the differences in pressure in the formation and the well) on at least one occasion. On another occasion they "lost hole" (the drill got stuck, apparently as a result of the heavy mud load and small drill) and had to abandon the drill and part of the drill string. To continue, it was necessary to seal off a 1000-foot section of the well with a cement plug and then drill a "side track" hole to circumvent the section. Finally, after a period of intense debate between the company's geologists and drilling engineers, the company made the decision to abandon the well.[53] The company then issued its cryptic news release.

Excessive formation pressures appear to be a standard condition in many parts of the Delta and Beaufort Sea. According to *Oilweek* much of Imperial's acreage is involved:

In the Mackenzie Delta, north of a roughly east-northeast trending major fault zone, tough geopressure zones must be expected. On the seaward or northward side of the fault a major downward movement of sediments has occurred and the corresponding sedimentary section (i.e. the section which is younger than the initial fault movement) is accordingly very much thicker than landward from this line. All of Imperial's finds and most of the wells drilled, drilling or planned in the Delta region lie on the "high pressure side" of this major feature.[54]

More than two years later the petroleum magazine indicated that the problem was a continuing one. "Geopressures continue to haunt Delta drillers with spectacular kicks putting BOPs to severe tests, and solid thrusts that threaten to lift conductor pipes complete with frozen surrounds, which was rumoured as an Immerk possibility."[55]

Many questions have been asked about the occurrence and nature of geopressure zones in the Richards Island-Beaufort Sea area, but few answers are available. It appears that Imperial hit abnormal pressure zones in a number of wells close to or below 10,000 feet, including probably Taglu, Hooper, Langley, and Adgo. The pressures made drilling more difficult but apparently were well within the range of the drilling equipment.

The problem of excessive geopressure has been encountered by several companies since Imperial abandoned Immerk. It was reported that "Sun Oil lost Pelly Island this season and is fighting pressures on its other man-made island location at Unark." Shell Canada was also reported to have had to abandon a land well, Unipkat 22, and Chevon was fighting excessive pressures at its Upluk well. The review of the problem by *Oilweek* stated that "Delta drillers are learning to recognize the high pressure wells and a belief is growing that they usually are water bearing and poor hydrocarbon prospects." But obviously *Oilweek's* headline, "High pressures tamed in Delta," was too optimistic for the article which said that knowledge of the occurrence of structures with excessive geopressures is still so inadequate that at least two wells have been abandoned during the past year and a number of others were in trouble because of them.[56]

Immerk was the first offshore well, drilled in the fall of 1973. By the second anniversary of the announcement of its abandonment, almost a dozen more will have been drilled in shallow water and two offshore drillships will be waiting for the spring breakup. One of *Oilweek's* articles on geopressure zones wisely refers to the need for time to gain experience. Considering the problem of actual or possible encounter with geopressure zones, it makes sense to proceed with a drilling programme on a step-by-step basis, allowing each subsequent hole to be better planned and drilled than the preceding one. But this will hardly be the case if a rush is underway — the normal situation with every play in western Canada. Hopefully, prohibitive costs — financial, environmental and social — in the frontier areas will impose a more cautious approach for the common benefit of all interests in the North.

References and Notes

1 The term "beyond compensation" was used by DOE officials to describe the consequences of a major oil spill in Hudson Bay and Lancaster Sound. See Chapters 6 and 7.

2 The blowout rate in U.S. outer continental shelf areas (presumably Louisiana and California) has been approximately one for each 500 wells drilled according to *Energy under the Oceans* (Norman, Oklahoma: University of Oklahoma Press, 1973).

3 DINA, Draft Memorandum to Cabinet, May 1973. See Appendix 1.

4 Johnstone, R.D., *Drilling in Cook Inlet*, in *Proceedings of Northern Canada Offshore Drilling Meeting*, December 1972.

5 See Appendix 6, for a detailed discussion of technological capability of cleaning up an oil spill in Arctic waters.

6 This section is based largely on a paper presented by Percy *(Arctic marine ecosystems and oil pollution)* at the Arctic Ecology Conference in Ottawa in September 1975 and Appendix 6.

7 Ramsier, R.O., *Possible fate of oil in the Arctic Basin*, paper presented at the First World Congress on Water Resources, September 1973, Chicago.

8 Percy, see n. 6.

9 The blowout at Santa Barbara and the *Torrey Canyon* spill resulted in a great many published accounts which widely informed the public on the fatal effects a crude-oil spill had on birds.

10 Percy, see n. 6.

11 Barry, T.W., *Likely effects of oil in the Canadian Arctic, Marine Pollution Bulletin,* Arctic issue, 1970.

12 Slaney, F.F., *Environmental impact assessment, Immerk artificial island construction, Mackenzie Bay, N.W.T.,* Imperial Oil, 1972.

13 Barry, see n. 11.

14 Barry, see n. 11.

15 Tuck, L.M., *The murres: Their distribution, populations and biology, a study of the Genus Uria,* Canadian Wildlife Service, 1960.

16 Barry, see n. 11.

17 Godfrey, W.E., *The birds of Canada* (Ottawa: National Museum of Canada, Biological series 73, 1966), and Snyder, L.L., *Arctic birds of Canada* (Ottawa: Information Canada, 1957).

18 Warner, R.E., *Environmental effects of oil pollution in Canada: An evaluation of problem and research needs,* a brief prepared for the Canadian Wildlife Service, 1969.

19 Department of Transport Task Force Report, vol. 1.

20 Brownell, R.L., Jr., *Whales, dolphins and oil pollution,* in *Ecological and oceanographic survey of the Santa Barbara Channel oil spill, 1969, 1970* (Berkeley: University of Southern California, Allan Bancock Foundation, 1973).

21 Smith, T.G., *Population dynamics of the ringed seal in the Canadian eastern Arctic,* (Ottawa: Fisheries Research Board of Canada Bulletin 18, 1973). See also Appendix 5.

22 See also Chapter 9.

23 Percy, see n. 6.

24 Council on Environmental Quality, *OCS oil and gas — an environmental assessment:* (Washington, D.C.: Superintendent of Documents, U.S. Government Printing Office, 1974).

25 Kendell, R.E. et al., *Movements, distribution, population and food habits of fish in the Western Coastal Beaufort Sea, 1974,* also Galbraith, D. and Fraser, D.C., *Movements, distribution, population and food habits of fish in the Eastern Coastal Beaufort Sea, 1974,* also Percy, R., Eddy, W., and Munro, D., *Anadromous and freshwater fish of the Outer Mackenzie Delta* (Ottawa: DOE Interim Reports of the Beaufort Sea Project, 1974).

26 Knowledge of fish populations throughout the Arctic is at a similar rudimentary state. See Sprague, J.B., *Aquatic resources in the Canadian North: Knowledge, dangers and research needs,* in Pimlott, D., et al., *Arctic alternatives* (Ottawa: Canadian Arctic Resources Committee, 1973).

27 Kendel, see n. 25.

28 Macdonald, C.B., and Leviw, C.P., *Geomorphic and sedimentologic processes of rivers and coast, Yukon coastal plain* (Ottawa: DINA, Task Force on Northern Oil Development, 1973).

29 Percy, J.A., *Effects of crude oil on Arctic marine ecosystems* (Ottawa: DOE, interim report of the Beaufort Sea Project, Study B6b, 1974).

30 Chronic spills are discussed in Chapter 4.

31 Council on Environmental Quality, see n. 24.

32 Climatic studies for the Beaufort Sea, which were conducted for the Arctic Petroleum Operators' Association by the Institute of Storm Research in Houston, Texas had a theoretical base because actual data for offshore areas were not available; see Chapter 9.

33 Thobe, S., *Storms are a major cause of mobile rig mishaps offshore, Offshore*, 5 June 1974.

34 Roots, E.F., *Sea ice and icebergs* (Ottawa: DINA, December 1972).

35 Rowsell, K.A., *Notes on some major technical problems* (Ottawa: DINA, December 1972).

36 Roots, E.F., *The sea floor and below*, (Ottawa: DINA, December 1972); see also Shearer, J.M., et al., *Submarine pingoes in the Beaufort Sea, Science*, November 1971.

37 McKay, G.A., et al., *A climatic perspective of tundra areas*, in *Proceedings of Conference on Productivity and Conservation in Northern Circumpolar Lands*. (Morges, Switzerland: IUCN, 1970).

38 Burns, B.M., *The climate of the Mackenzie Valley — Beaufort Sea* (Ottawa: DOE, Climatological studies no. 24, 1973).

39 DPW, Engineering Programmes Branch, *Beaufort Sea storm — Investigation of effects in the Mackenzie Delta* (Ottawa: DOE, November 1971).

40 Wilson, H.P., *Study of the Beaufort Sea storm of September 1970*, unpublished report, Canadian Arctic Weather Control, Edmonton 1971.

41 Burns, see n. 38, vol. 2, 1974.

42 Burns, see n. 38, vol. 1, 1974.

43 *Oilweek, Geopressure zones throw challenge to drillers in three Canadian plays*, 4 December 1961.

44 *Oilweek, Occurrence of gas hydrates in Delta creates new problems for wildcatters*, 3 June 1974.

45 Deju, R.A., *A worldwide look at the occurrence of high fluid pressures in petroliferous basins*, Wright State University, Department of Geology, 1973.

46 *Oilweek*, see n. 44.

47 Deju, see n. 45.

48 Levorsen, I.A., *Geology of petroleum* (San Francisco: Freeman, 1974).

49 Deju, see n. 45.

50 Council on Environmental Quality, *OCS Oil and Gas — an environmental assessment* (Washington, D.C.: Superintendent of Documents, U.S. Government Printing Office, 1974).

51 Kash, D.E., et al., *Energy under the oceans: A technological assessment of Outer Continental Shelf oil and gas operations* (Norman, Oklahoma: University of Oklahoma Press, 1973).

52 *Oilweek, Mud weight control on the Delta*, 17 March 1975.

Although it is evident that abnormal formations are a hazard to drilling, and particularly to the drilling of wildcat wells, it appears that Imperial has not been very frank with DINA about it. In fact, there is evidence which suggests that Imperial may have deliberately misrepresented the nature of the problem in making application to drill Immerk B-48. It is important that the matter be clarified: Immerk was the first offshore well, and as discussed in Chapter 3, it established a precedent which was to be used to advantage in seeking approval to construct and drill from other artificial islands. If the data provided were in fact false or misleading it would raise doubts about the veracity of the information provided by Imperial on its entire construction and drilling programme.

The evidence is as follows: In support of its application to drill, Imperial submitted an "Arctic Well Contingency Plan — Immerk B-48." Section 1330 dealt with preplanning for emergencies and contained a paragraph on formation pressures. The first sentence stated what Imperial expected to encounter in drilling Immerk: "Analysis of available velocity data indicates that pressures above a depth of 10,000 feet will be normal and equivalent to a salt water gradient." However, in March 1975 Mr. R.N. Daw of Imperial Oil presented a paper to the Canadian Society of Petroleum Geologists on the detection and control of overpressure. *Oilweek* reported on his talk in an article entitled "Mud weight control in the Delta," and included two sets of graphs from his paper. One, "Immerk velocity plot," showed that the company anticipated overpressure at 7600 feet. According to *Oilweek,* "Seismic indicated overpressure at Immerk and its depth and severity were predicted from a velocity plot using background curves from the Gulf Coast. Imperial took into consideration that these curves tend to underrate Beaufort conditions. The overpressure was estimated at 7600 feet and was expected to be severe, calling for at least 16 lbs/gal pore pressure equivalent."

53 Information on problems encountered in drilling Immerk was obtained in discussion with members of government and industry.

54 *Oilweek*, n. 43.

55 *Oilweek, Accelerated interest in Beaufort Sea as 6 more man-made islands proposed,* 4 March 1974.

56 *Oilweek, High pressures tamed in Delta,* 3 March 1975, also *Oilweek,* see n. 43.

Chapter 9

Research and Environmental Assessment

There is a remarkable similarity in the behaviour of government and industry in preparing for the Beaufort Sea Project and in preparing for the construction of the Mackenzie Valley Highway. The story of the highway is well known. Soon after the Prime Minister announced in April 1972 that highway construction would begin immediately, it became generally known that no social or environmental studies had been conducted by the Government, nor had any detailed engineering surveys been done prior to the announcement. In the Beaufort Sea, the Government had been granting exploration permits since 1961. Under the terms of some permits, the operators were obliged to drill by 1974 or 1975, although the Government had never undertaken any extensive environmental work in the area.[1]

By 1973, the petroleum industry, on the other hand, had spent more than $2 million on research on ways of coping with the harsh conditions of the Beaufort Sea while exploring for oil and gas. More than $500,000 was invested in engineering studies alone through the Arctic Petroleum Operators' Association (APOA).[2] The industry had also spent, and continues to spend, millions of dollars on seismic studies to identify potentially productive areas. But by the time it was confronted with applications to drill, the Government had taken virtually no action to obtain the basic information on various forms of marine life, currents, and ice which it would need in order to judge the practicality of exploring for oil and gas, to regulate exploration, to assess the potential environmental impact of the project, or to assess contingency plans for dealing with oil spills from blowouts.

In April 1973, the Oil and Minerals Division of DINA completed a 109-page position paper on offshore drilling. The document, which has never been released publicly, states that:

The primary basic information relating to the physical environment required to formulate design criteria for Arctic operations, includes basic meteorology such as temperature, wind speed and direction, sea state data such as average and maximum wave heights, basic ice data such as ice thickness and strength, ice forces, distributions and movement, the geometry, composition and strength of ice islands and pressure ridges, sea bottom data relating to soil conditions and the frequency and distribution of ice scouring.

Primary basic information required to identify the ecology in the region of the Beaufort Sea includes baseline data on the quality of the sea, air and land inventories of the type and amount of sea life in the region, data on the extent of the marine resources utilization of the area, and impact studies to relate the ecology of the Beaufort Sea to an Arctic offshore drilling system.[3]

Thus the position paper seemed to recognize the need for research. But in another section, the report suggested that a great deal of knowledge had been gained already through a number of on-going projects; the Polar Continental Shelf Project, the Arctic Ice Dynamics Joint Experiment (AIDJEX), and the work of a number of other agencies including the U.S. Naval Research Lab, U.S. Army Cold Regions Research, the University of Alaska, the Arctic Institute of North America, and the U.S. Department of the Interior. The view that existing knowledge was sufficient prevailed, because the draft memorandum to Cabinet on the subject of offshore drilling[4] did not suggest that any further studies were required prior to granting final approval for drilling. But before the matter was to be considered by Cabinet on 31 July 1973, the Department of the Environment (DOE) became involved; as a result, Cabinet did give some consideration to the need for research prior to drilling.

The Politics of the Beaufort Sea Project

In giving DINA approval in principle to license offshore operations, Cabinet stipulated that DOE would conduct a crash research programme in the Beaufort Sea. The research initially was estimated to cost approximately $5.5 million to be paid for by industry. It was to be completed in 18 months, by the beginning of the drilling season in 1975 when both the Beaufort Sea Task Force (BSTF) and Hunt International Petroleum Co. expected to begin exploratory drilling. DOE was to compile the list of studies to be conducted before the drilling programme got underway.[5]

The studies were to be managed by a new inter-departmental group, the Arctic Waters Oil and Gas Advisory Committee (AWOGAC), which would also regulate the environmental aspects of offshore drilling projects. On 20 September, Mr. Chrétien advised BSTF and Hunt that approval in principle was granted to begin drilling in 1975. The approval conditions required operators to finance the research programme which was to be directed by DOE. The companies, however, objected to this requirement and the AWOGAC met on 5 October 1973 to try to resolve the problems. Members of both the BSTF and Hunt attended.

The companies reacted strongly against the proposal for a number of reasons. Representatives of Hunt and BSTF argued that many companies eventually would be operating in the area, and that newcomers would receive the benefit of the initial studies without having to pay for them. They also argued that basic or baseline environmental work should be paid for by government and not by industry. They may also have feared that if the studies were conducted, the results might bring a complete halt to drilling. Treasury Board had originally requested DOE to submit the final financial arrangements for the study package by 16 October, but the oil companies did not respond by the deadline. DOE and the companies finally agreed that the cost of the studies would be shared equally by government and industry. Under this arrangement, APOA was to represent the industry on the coordination committee for the Beaufort Sea Project.

During the lengthy negotiations between the Government and the companies, DOE and DINA clashed on the nature and direction of the proposed studies. Apparently DINA insisted on its right to assert more direct control over the planning and conduct of the studies. In the meantime, DOE had already appointed a manager for the research programme and established his headquarters at Victoria, B.C. A meeting was held between the two departments on 24 January 1974 to resolve the problem; subsequent events indicated that DOE maintained its control of the research programme.

The matter finally reached Treasury Board on 15 February. Government and the companies agreed that DOE would need $2.2 million to finance its half of the study programme, but because the funds were not included in the department's annual budgetary estimates, they required Treasury Board approval. While DOE officials expected that Treasury Board might cut back the funds slightly, they were unprepared for the Board's subsequent decision to reject completely DOE's request for new funds. The Treasury Board decision was not made public.[6]

The Board rejected DOE's request on a Friday. The following Monday, APOA issued a news release announcing that a $2 million environmental impact study of the Beaufort Sea would be carried out as a "joint undertaking of APOA and the federal Government."[7] The statement made no mention of the earlier Treasury Board decision on DOE's research project. Nor did it mention that the original environmental assessment package included $4 to $5 million worth of research studies. Nor did it explain why APOA had waited until three days after the Treasury Board decision to announce a programme which had been agreed to a month before.

Ironically, the APOA release ended with the statement that "perhaps now the obstructionists with the pronouncements which mislead the public will be prepared to acknowledge that industry and government have been taking direct constructive steps to obtain the pertinent information prerequisite to drilling."[8] The statement also revealed the industry's anxiety to begin drilling in 1975, saying that "enough of these studies will be completed by the end of this year to provide government with sufficient information to issue drilling authorities in the Beaufort for the 1975 summer season." The Treasury Board decision was followed by two weeks of intense discussion between DOE, DINA, and the industry. On 6 March, in its first public statement on drilling in the Beaufort Sea, DINA announced that the industry would pay for all the environmental studies at an estimated cost of $4.1 million and that drilling would not begin until 1976.[9] Although the statement noted that the delay in drilling would provide "over two years" for environmental studies, DOE officials maintained that plans still called for all the studies to be completed by July 1975 (16 months). At the same time, DINA also announced that Hunt and Dome Petroleum had been given permission to begin immediate construction of their respective drilling systems.

But in spite of the announcement by the Minister, the industry had still not submitted its last argument, subsequently described by the Beaufort Sea project manager as follows:

After some months of negotiation, industry agreed to support 21 of the 29 Beaufort Sea Studies, to the sum of $4,100,000. The balance of $1,200,000 required to fund the remaining 8 studies was to be funded from within the Department of the Environment. The agreement under which the industry was to transfer funds to government was signed on 15 May 1974. The historic agreement states that the government will coordinate the Beaufort Sea Studies and maintain cost control. Provision is made for an industry project manager to monitor progress on behalf of the oil operators and to jointly reallocate funds between studies with the government Project Manager. Resolution of differences is the responsibility of a senior government committee.[10]

One interesting result of the political manoeuvring was that DOE was able to hold its own in a jurisdictional argument with DINA. The original intention was to give AWOGAC the executive role, putting DINA in a very strong position because AWOGAC was created by DINA and is kept under tight rein by its chairman, who is a member of DINA; DINA wanted even more direct control over the project, possibly to the extent of having at least part of it brought under its Arctic Land Use Research programme. Neither of these objectives was achieved. The executive committee as finally constituted contains both DINA and industry representatives, but neither it nor the Beaufort Sea Project was subject directly to DINA during the operation of the project.

The final reports and recommendations of the project are made to the AWOGAC; it in turn reports to the Northern Natural Resources and Environment Branch of DINA, which will have the responsibility of implementing recommendations. It is a process that will warrant watching. The actions taken at the three stages will tell a great deal about the state of evolution of government policies on environmental protection.

Offshore Research by Industry, 1970-1975

Research by industry in the Beaufort Sea region began in 1970. Various studies have been conducted as joint programmes through APOA, whose reports are considered to be of a proprietary nature and therefore are closely guarded. At the Offshore Drilling Meeting, Dr. J. Woodward, chief of DINA's Oil and Minerals Division, stated that:

All projects undertaken, or being undertaken, by the Arctic Petroleum Operators' Association, numbering over 50, and accounting for expenditures of almost $3 million, have been allowed as "participation research projects." . . . A stipulation in respect of all research projects approved under the Regulations, is that they shall not be held confidential for more than five years after completion, or for a lesser period if the sponsor should agree . . . . With the exception of certain industry reports, including most seismic surveys, these are open in our office, upon selective approval, to those Government specialists with a need to know.[11]

In terms of public interest, it is very unsatisfactory that research reports related to the environment are classed confidential for five years after completion. This is particularly true for a highly hazardous area like the Beaufort Sea. Documents which are open "upon selective approval, to those government specialists with a need to know," can hardly be described as sufficiently accessible even for internal review purposes.

Of the 50 projects which have been allowed as participation research projects until the end of 1972, 29 appeared to relate either directly or indirectly to offshore drilling in the Beaufort Sea. The total cost of the projects was $2,633,422. Of this, 23% ($604,910) was spent on environmental studies, with the majority of that going on a single project (No. 43) which was described as an "Environmental impact assessment program — Mackenzie Delta."[12] The operator of the project was Imperial Oil, and the study itself was done by F.F. Slaney & Co. of Vancouver. The "position paper on offshore drilling" refers to Slaney & Co.'s research as "an ecological impact study of oil and gas exploratory drilling in the vicinity of the Southern Beaufort Sea for Imperial Oil Ltd." The study's objective is misrepresented by that description. Slaney prepared two reports for Imperial Oil. The first was an interim report entitled "Environmental field program, Taglu-Richards Island, Mackenzie Delta," described as studies "to provide a biophysical environmental basis for production facility impact assessment and to assist in planning development facilities and programs in a way that will minimize environmental effects. The current exploration program of Imperial Oil is not within the scope of the study." The letter of transmittal stated that "The report is a review of our approach to the problem of obtaining an adequate baseline inventory of resources and environmental quality. It also provides a preview of the type and extent of supporting environmental data that will accompany the environmental impact statement to be submitted at a later date. It is not an environmental impact assessment."[13]

The second report, also submitted in January 1973, is called "Environmental impact assessment, Immerk artificial island construction, Mackenzie Bay, N.W.T."[14] There are two points to be made about this impact assessment. First, the study was done concurrently with the first stage of construction of Immerk as a result of a recommendation made by the Land Use Advisory Committee. The application to construct Immerk was approved on the basis of another report, Imperial Oil Enterprises Limited Report No. IPRE-ZIME-71 (July 1971). James Woodford, who saw the report, described it this way: "It was only 19 pages long and was divided into four sections, including four pages on environmental impact. The report did not contain any new scientific information, any baseline data on the ecological conditions of the area, and did not list any special Imperial Oil studies on the impact of artificial islands on the environment."[15]

Second, as the title indicates, the report had nothing to do with exploratory drilling. Its sole purpose was to deal with the environmental impact of the island's construction. Slaney's letter of transmittal left no doubt on this point: "According to our terms of reference, both the study aspect of the project and the environmental statement have dealt specifically with the matter of construction of the artificial island, Immerk, along with directly associated activities. Eventual disposition or use of the island in any way, such as drilling for hydrocarbons, was beyond our terms of reference to consider."[16] A further study of beluga whales was also conducted by Slaney for Imperial during 1973.[17]

Clearly, none of the studies undertaken by industry up to the time that Cabinet gave approval in principle to offshore drilling had been designed to provide information on the ecological impact of oil and gas exploratory drilling or on oil spills in any part of the Beaufort Sea. Three APOA projects in particular, No. 12 ("All season exploratory drilling system — 0 to 200 feet of water"), No. 13 ("Seasonal drilling from a barge"), and No. 30 ("Beaufort Sea exploratory drilling system") are discussed at length in the "Position paper on offshore drilling." All three projects were essentially economic and engineering feasibility studies.

Even the briefest glance at these APOA projects raises many questions and underlines the need for such reports to be available for critical review. For example, the first part of Project No. 13 was a report on winds, waves, and storms in the southern Beaufort Sea by Dr. John Freeman of the Institute of Storm Research in Houston, Texas. The study was "a hindcast study of waves and wind conditions," and the position paper went on to state that "winds were calculated from weather maps and the waves calculated from those winds and fetches, since wind measures were practically nonexistent." In other words, the background data on wind velocity and wave characteristics were theoretical, not recorded. Does this provide an adequate basis for the design of an exploratory drilling system for the Beaufort Sea? The question should be answered by independent climatologists and meterologists who have been given the opportunity to make a critical review of the report.[18]

At the Northern Canada Offshore Drilling Meeting, and in the position paper on offshore drilling, it was emphasized that the two APOA reports had independently recommended essentially the same barge drilling system. But even to a layman, the questions which cry out for answers are, What were the criteria? What were the terms of reference given to the consultants who prepared the reports? How stringent were the financial guidelines? Is it possible that they could have outweighed environmental protection considerations in the development of a system for use in the Beaufort Sea? If the drilling barges which were recommended were adequate for operations in the Beaufort Sea, why was the concept of drill barges abandoned after it had been approved by Cabinet?

TABLE 1
BEAUFORT SEA RESEARCH PROGRAMMES (APOA)
1970-1975

Table 1a Ice Forces and Properties

Proj. No.	Oper.	P.C.*	Description	Cost
1	IOL	8	Nutcracker-Large Scale Ice Strength Tests	$166,790.00
9	IOL	5	Nutcracker Phase II	60,000.00
52	IOL	4	Measuring the Crushing Strength of Ice	141,500.00
25	IOL	8	Model Test Simulating Ice on Fixed Structure	59,000.00
16	IOL	5	Theoretical Analysis of Ice Failure	10,000.00
40	IOL	8	Eval. of Mech. Prop. of Saline Model Ice	19,800.00
41	AMOCO	9	Eval. of Prop. of Michel's Model Ice	9,000.00
50	Sun	3	Ice Thickness Measurement	27,900.00
54	IOL	2	Ice Geology-Southern Beaufort Sea	50,000.00
2	IOL	15	Properties of Sea Ice and Current Measurements	351,249.00
57	IOL	7	Effects of Ice Adhesion (or freeze) on a Conical Structure	110,000.00
66	IOL	4	Ice Crushing Tests 1973/74	115,000.00
68	IOL	4	Properties of Model Ice Ridges	25,000.00
75	IOL		Field Study of First Year Pressure Ridges	80,000.00

* P.C. Participating Companies

Table 1b Distribution and Movement of Ice

Proj. No.	Oper.	P.C.	Description	Cost ($)
14	AMOCO	10	Summer Ice Reconn. Beaufort Sea	20,000.00
31	IOL	4	Aerial Reconn. of Ice Beaufort Sea	6,000.00
46	Sun	1	Ice Reconn. April 1972 Beaufort Sea	12,600.00
51	IOL	3	Ice Movement in the Beaufort Sea 1972-73	102,500.00
53	IOL	2	Count of Ice Islands in Beaufort Sea 1972	10,000.00
33	IOL	4	Landfast Ice Movement-Mackenzie Delta	90,000.00
60	Gulf	10	1973 Summer Ice Study Beaufort Sea	88,336.00
67	IOL	3	Landfast Ice Movement Beaufort Sea 1973/74	105,000.00

Table 1c Ocean Bottom Studies

Proj. No.	Oper.	P.C.	Description	Cost ($)
19	Gulf	12	Sea Bottom Scouring-Analysis of Records	16,329.00
32	Gulf	8	Sea Bottom Scouring 1971-72	22,000.00
3	IOP	14	Ocean Floor Sampling	425,686.00
4	Gulf	9	Geological Analysis of Ocean Floor Samples	11,757.00
23	Gulf	1	Soil Strength of Beaufort Sea Samples	3,000.00
59			1973 Beaufort Sea Scouring Study Phase III	
69	IOL	6	Analytic Study of Ice Scour	30,000.00

Table 1d Environmental Studies

Proj. No.	Oper.	P.C.	Description	Cost ($)
11	IOL	3	Ornithological Study — Mackenzie Delta	7,660.00
28	IOL	7	Biological Effects of Oil in Arctic Seawater	17,250.00
55	APOA	18	Arctic Environmental	273,511.00
73	IOL	17	Research Program on Pollution from Drilling Fluids	130,700.00
76	IOL		Summer Environmental Studies East Mackenzie Bay, Mackenzie Delta, N.W.T.	451,000.00

Table 1e Engineering Studies

Proj. No.	Oper.	P.C.	Description	Stat.	Cost ($)
13	Elf	11	Arctic Drilling Barge Study	C	134,601.00
30	Gulf	11	Beaufort Sea Exploratory Drilling Systems	C	33,000.00
12	IOL	12	All Seasons Exploratory Drilling System — 0 to 200 ft.	C	150,900.000
39	IOL	4	Submarine Pipeline Study - Offshore Mackenzie Delta	IP	75,000.00
36	Gulf	9	Ice Island Destruction — Beaufort Sea	C	20,000.00
65	IOL	6	Small Prototype Cone Test	C	238,269.00
77	IOL	4	Modelling of Small Cone Prototype Tests	IP	40,600.00

Table 1f Other Relevant Projects

Proj. No.	Oper.	P.C.	Description	Stat.	Cost
63	APOA		Arctic Institute of North America's Beaufort Sea Symposium	C	$ 12,000.00
70	IOL	4	Wind/Wave Hindcast, Canadian Beaufort Sea	IP	49,800.00

The offshore studies commissioned by APOA (apart from those conducted as part of the Beaufort Sea environmental program) from 1970 to the present time are shown in Table 1. The studies have been subdivided by class to show how the research effort was allocated during the period. The table makes it clear that the research was almost totally related to acquiring knowledge that was relevant to the conduct of drilling operations, and it was not until DOE made a stand in 1973 that member companies of APOA felt compelled to make a major investment in research related to the protection of the environment.

The Beaufort Sea Project

The Beaufort Sea research project raises many questions about scientific research and its application in the assessment of potential environmental effects of major projects such as offshore drilling in the Beaufort Sea. Two important queries are, How valuable is the knowledge gained through crash "catch-up" programmes of baseline ecological research? Is it valid and realistic to prepare environmental impact assessments for areas with highly variable environments, based on research programmes done in only one or two seasons?

The Beaufort Sea Project originally consisted of 29 studies, but since the signing of the agreement between government and industry on 15 May 1974, three other government-funded studies have been added.[19] The plan calls for the 32 studies to be completed and synthesized into six overview reports which will emphasize the offshore exploration and related activities in the Beaufort Sea. A final summary document will be prepared from both the individual and the overview reports. The advance billing for this report is that it will be "a concise, nontechnical summary of the major conclusions and recommendations" outlined in the studies and overview reports.[20] The report will focus on major findings and alternatives stressed in the previous studies, and will attempt to relate the environmental data to the environmental risks. It will also suggest the kinds of conditions to be met when drilling in deepwater areas of the Beaufort Sea. December 1975 is the projected completion date for the report, and at that time hopefully the information will be available for both public and government assessment and review.

According to Dr. Al Milne, the manager of the project, the Beaufort Sea research programme was designed with three principal objectives:

i) to increase our knowledge of the southeast Beaufort Sea so that Arctic Waters Oil and Gas Advisory Committee, and ultimately Cabinet, can rule on petroleum industry applications for offshore exploratory drilling:

ii) to define the seasonal and geographical sensitivities of the Beaufort Sea so that the impact of future drilling activities in the region will be minimized;

iii) to provide a design for a weather and sea-ice prediction system so that drillships and platforms can do exploratory drillings with a minimum of hazard to their personnel and to the environment.[21]

Milne stated at a meeting in October 1974 that "the scope of the project is defined by three constraints. It is concerned with the southeastern part of the Beaufort Sea, it must be completed by December 1975, and it is confined to the impact of the exploratory drilling phase of offshore oil operations."[22] He went on to discuss the threat posed by the blowout of an oil well and stated that although "the probability of a blowout is remote . . . as the number of exploratory holes remains to plague us, a blowout at some future date is almost inevitable." In the face of

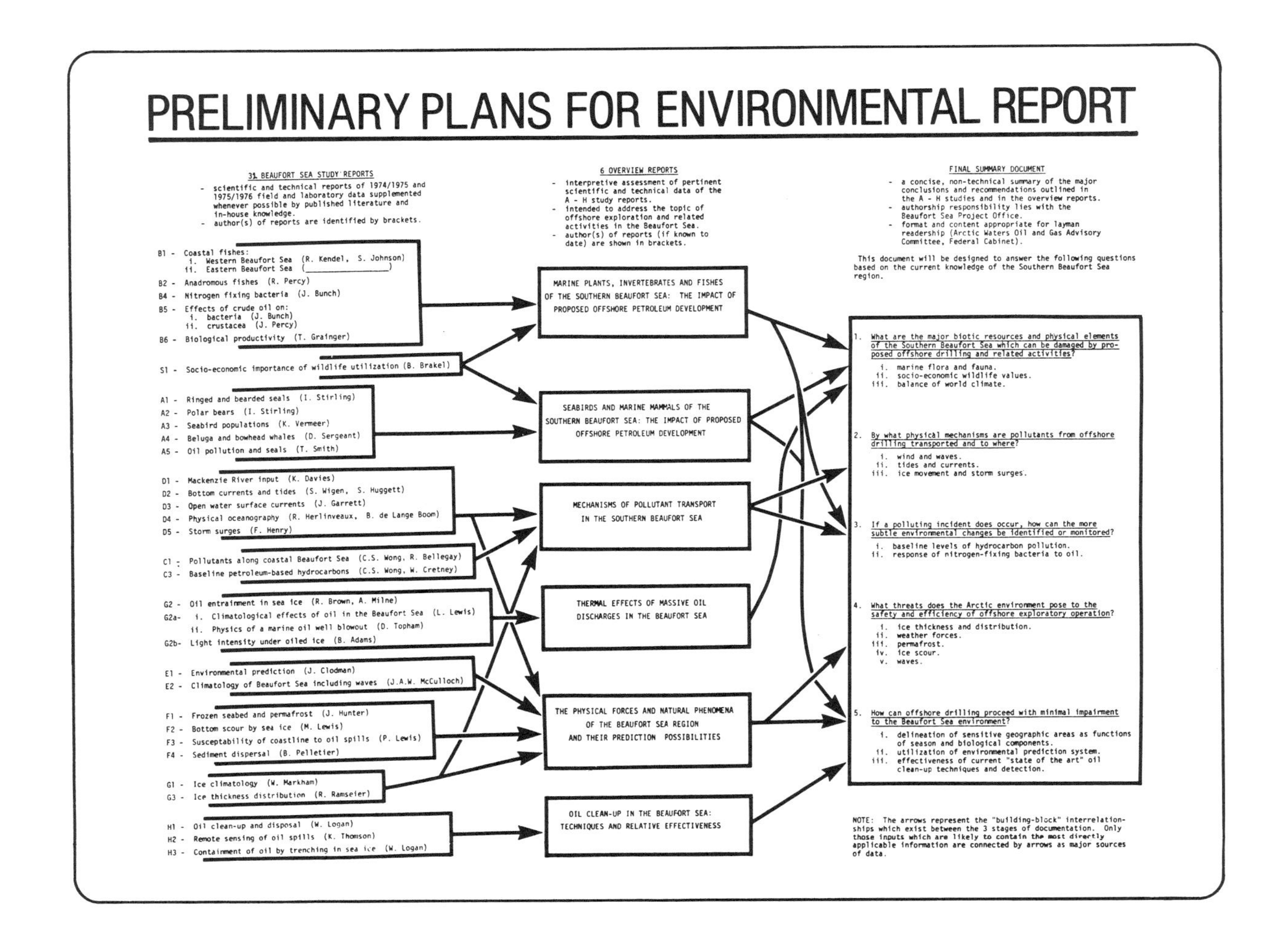
PRELIMINARY PLANS FOR ENVIRONMENTAL REPORT
31 BEAUFORT SEA STUDY REPORTS
- scientific and technical reports of 1974/1975 and 1975/1976 field and laboratory data supplemented whenever possible by published literature and in-house knowledge.
- author(s) of reports are identified by brackets.
B1 - Coastal fishes:
i. Western Beaufort Sea (R. Kendel, S. Johnson)
ii. Eastern Beaufort Sea (______________)
B2 - Anadromous fishes (R. Percy)
B4 - Nitrogen fixing bacteria (J. Bunch)
B5 - Effects of crude oil on:
i. bacteria (J. Bunch)
ii. crustacea (J. Percy)
B6 - Biological productivity (T. Grainger)
S1 - Socio-economic importance of wildlife utilization (B. Brakel)
A1 - Ringed and bearded seals (I. Stirling)
A2 - Polar bears (I. Stirling)
A3 - Seabird populations (K. Vermeer)
A4 - Beluga and bowhead whales (D. Sergeant)
A5 - Oil pollution and seals (T. Smith)
D1 - Mackenzie River input (K. Davies)
D2 - Bottom currents and tides (S. Wigen, S. Huggett)
D3 - Open water surface currents (J. Garrett)
D4 - Physical oceanography (R. Herlinveaux, B. de Lange Boom)
D5 - Storm surges (F. Henry)
C1 - Pollutants along coastal Beaufort Sea (C.S. Wong, R. Bellegay)
C3 - Baseline petroleum-based hydrocarbons (C.S. Wong, W. Cretney)
G2 - Oil entrainment in sea ice (R. Brown, A. Milne)
G2a- i. Climatological effects of oil in the Beaufort Sea (L. Lewis)
ii. Physics of a marine oil well blowout (D. Topham)
G2b- Light intensity under oiled ice (B. Adams)
E1 - Environmental prediction (J. Clodman)
E2 - Climatology of Beaufort Sea including waves (J.A.W. McCulloch)
F1 - Frozen seabed and permafrost (J. Hunter)
F2 - Bottom scour by sea ice (M. Lewis)
F3 - Susceptability of coastline to oil spills (P. Lewis)
F4 - Sediment dispersal (B. Pelletier)
G1 - Ice climatology (W. Markham)
G3 - Ice thickness distribution (R. Ramseier)
H1 - Oil clean-up and disposal (W. Logan)
H2 - Remote sensing of oil spills (K. Thomson)
H3 - Containment of oil by trenching in sea ice (W. Logan)
6 OVERVIEW REPORTS
- interpretive assessment of pertinent scientific and technical data of the A - H study reports.
- intended to address the topic of offshore exploration and related activities in the Beaufort Sea.
- author(s) of reports (if known to date) are shown in brackets.
MARINE PLANTS, INVERTEBRATES AND FISHES OF THE SOUTHERN BEAUFORT SEA: THE IMPACT OF PROPOSED OFFSHORE PETROLEUM DEVELOPMENT
SEABIRDS AND MARINE MAMMALS OF THE SOUTHERN BEAUFORT SEA: THE IMPACT OF PROPOSED OFFSHORE PETROLEUM DEVELOPMENT
MECHANISMS OF POLLUTANT TRANSPORT IN THE SOUTHERN BEAUFORT SEA
THERMAL EFFECTS OF MASSIVE OIL DISCHARGES IN THE BEAUFORT SEA
THE PHYSICAL FORCES AND NATURAL PHENOMENA OF THE BEAUFORT SEA REGION AND THEIR PREDICTION POSSIBILITIES
OIL CLEAN-UP IN THE BEAUFORT SEA: TECHNIQUES AND RELATIVE EFFECTIVENESS
FINAL SUMMARY DOCUMENT
- a concise, non-technical summary of the major conclusions and recommendations outlined in the A - H studies and in the overview reports.
- authorship responsibility lies with the Beaufort Sea Project Office.
- format and content appropriate for layman readership (Arctic Waters Oil and Gas Advisory Committee, Federal Cabinet).
This document will be designed to answer the following questions based on the current knowledge of the Southern Beaufort Sea region.
1. What are the major biotic resources and physical elements of the Southern Beaufort Sea which can be damaged by proposed offshore drilling and related activities?
i. marine flora and fauna.
ii. socio-economic wildlife values.
iii. balance of world climate.
2. By what physical mechanisms are pollutants from offshore drilling transported and to where?
i. wind and waves.
ii. tides and currents.
iii. ice movement and storm surges.
3. If a polluting incident does occur, how can the more subtle environmental changes be identified or monitored?
i. baseline levels of hydrocarbon pollution.
ii. response of nitrogen-fixing bacteria to oil.
4. What threats does the Arctic environment pose to the safety and efficiency of offshore exploratory operation?
i. ice thickness and distribution.
ii. weather forces.
iii. permafrost.
iv. ice scour.
v. waves.
5. How can offshore drilling proceed with minimal impairment to the Beaufort Sea environment?
i. delineation of sensitive geographic areas as functions of season and biological components.
ii. utilization of environmental prediction system.
iii. effectiveness of current "state of the art" oil clean-up techniques and detection.
NOTE: The arrows represent the "building-block" interrelationships which exist between the 3 stages of documentation. Only those inputs which are likely to contain the most directly applicable information are connected by arrows as major sources of data.

this risk, he then listed the questions that the Beaufort Sea studies have been designed to answer. These are also the questions to be dealt with in the summary report:

— *What can be damaged in the environment of the Beaufort Sea? Examples are fisheries, marine mammals, birds which depend on the sea, and possibly the world's climate.*

— *How are pollutants transported and to where? Important mechanisms about which we know little include the ocean currents, sea-ice movements and storm surges.*

— *If damage does occur, how can changes be identified? For example, the detection of change in a living system requires knowledge of the system prior to the commencement of offshore drilling.*

— *How can exploratory drilling occur with minimum damage to the environment? We need to make knowledgeable choices of time of year and location where drilling will be permitted.*

— *What threats does the environment pose to the safety of offshore drilling? The critical factor here is the validity of the engineering data used for the design of drilling systems.*

At another point in his paper Milne assessed the relation of studies on environmental threats and the drilling systems: "None of the studies referred to is directly related to the engineering integrity of offshore platforms or equipment. Rather, the concern of the Department of the Environment is with the effect of the intrusion of drilling operations on the environment, and with whether or not the environmental data used by the systems and structural engineers in the design of offshore drillships and platforms are valid."[23]

The kinds of problems which arise in Arctic marine research programmes were dramatically illustrated by ice conditions in the Beaufort Sea in 1974. Intensive research was planned for an open-water season of 2.5 months. As it turned out, 1974 was the year when summer almost did not come to the Beaufort Sea. The research vessels, the *MV Theta* and *MV Pandora* were escorted by the icebreaker *Camsell* through the Bering Strait and into the Beaufort Sea. They did not arrive at Herschel Island until 9 August, three weeks later than expected.

The heavy ice conditions forced cancellation of some studies and allowed only limited progress on a number of others. Little progress was made on "Near bottom currents and offshore tides" (D2), and "Open water surface currents" (D3). Placing offshore instruments for automatic recording of data on currents and tides was part of the plan for D2; five were placed in November 1973 and 19 in May 1974. When the work to recover the instruments was started in August only one location was ice-free. Recovery attempts here proved unsuccessful, but fortunately two weeks later the instruments were found during the course of a helicopter flight for another project. Recovery from another ice-infested location was the only other successful venture. These instruments were retrieved six miles away from where they were supposed to be, apparently having been displaced by ice. Further recovery operations of instruments placed in 1974 constituted a major part of the 1975 field programme. Many of the gauges were not located during the season. The open-water surface currents study (D3) was forced to abandon its original plan and restrict efforts to nearshore currents in Mackenzie and Kugmallit Bays and in the vicinity of Atkinson Point.

Much of the work on the use of the Beaufort Sea by waterfowl and sea birds was restricted to a relatively small patch of open water north of the Tuktoyaktuk Peninsula. At no time did the area of open water extend to the sites which will be occupied by the drillships in 1976.

In January 1975, a Beaufort Sea Investigators' conference was held in Calgary to review progress made during the summer on the various research projects. Several of the interim reports pointed to the need for further studies and two in particular made concrete suggestions. One, on fish, stated the case this way:

Information on life history, movements, spawning and overwintering areas is lacking for all but a few species. With the exception of least cisco, Arctic cisco, Arctic char and inconnu, the information available can only be applied to the four-month ice free season. Very little is known about fisheries resources of the true marine area or the areas of permanent ice cover. The apparent variability of species abundance and distribution from year to year suggests that a five year programme would not be unreasonable. Due to limitations imposed on investigator by climate, light and ice conditions, work in this area is very much restricted. Existing equipment for winter survey work is inadequate to provide meaningful results.[24]

Another report, on the biological productivity of the Beaufort Sea, said succinctly:

Most troublesome at present is the assessment of the immense variability which living systems present. Add to this the need for long-term investigations to take into account non-biological environmental variations and such things as potential cumulative effects of the many kinds of impact which contemporary exploitation of marine resources now faces us with, and we are left in a quandry which a single year's work will not get us out of, either in this or in any other ecosystem study. We must develop a knowledge of relationships in the system and we must subject the elements of it to long-term impact studies. All of this will take far more time than is being given for this Beaufort Sea study.[25]

The question of time is always raised when the value of the Beaufort Sea Project is discussed. It was difficult not to react negatively at first to the concept of a one-year crash research programme. The case was stated this way: "I believe that a one-year crash research project for the Beaufort Sea constitutes a waste of public funds. Environmental conditions vary so much from year to year that a three-year programme would be the minimum required to produce a reasonable understanding of what could be done to handle and minimize the effect of oil spills."[26]

It appears that the potential value of the Beaufort Sea Project is more positive now, in October 1975, than it was in January 1974, although some skepticism is warranted at least until the results of the research are presented and incorporated into an environmental impact assessment. There is no doubt that the results of a three- or a five-year programme would hold more meaning for the assessment process than those obtained from an 18-month study. However, it is possible that a crash research programme may be better than no research at all. A great deal will depend on how the results are used in the assessment process. There is a tendency in some areas of government and in the petroleum industry to make exaggerated claims about the adequacy of the Beaufort Sea Project. If this tendency is continued it could affect the validity of the environmental impact assessment.

The behaviour of oil in ice and methods of cleaning up oil spills are two areas in which it will be difficult to keep research results from being over-interpreted and over-extended.[27] The Beaufort Sea research on oil is virtually the first research of any consequence concerning oil in Arctic conditions. The amounts of oil spilled and the areas involved in the experiments are minuscule compared to what would be involved if an exploratory well blew out and was uncontrolled for even a month.[28] It is hoped that the engineers and scientists involved will not make quantum leaps when drawing conclusions from "square one" data.[29] But there will surely be some knowledge gained which can be applied in a way which will mitigate the impact of some of the phases of the exploration processes. As noted earlier, it will be informative to watch how recommendations are implemented.

The greatest contributions of the Beaufort Sea Project will be indirect: the identification of knowledge gaps about the Beaufort Sea environment; clarification of the nature and kinds of research which should be undertaken in other Arctic offshore areas; development of a clearer understanding of environmental impact assessment processes and how the results can be applied in Arctic marine areas; and the development of public understanding of the inadequacy of knowledge about the behaviour and effect of oil in Arctic waters, of the lack of technological capacity to deal with spills which occur in pack ice, and of the dangers of offshore operations to the environment.

Environmental Impact Assessment and Offshore Drilling

As indicated by the case histories discussed in this book, a great deal of time and money is being spent in the North on environmental impact assessments or on background research connected with them. It is worth asking whether or not assessment is likely to make an important contribution to the protection of Arctic marine environments. Environmental impact assessment (EIA) was the subject of both a workshop and a national conference in Winnipeg in late 1973.[30] Both meetings drew heavily on experiences and insights that had been gained in the N.W.T. and the Yukon. The workshop was organized by the Environmental Protection Board (EPB). Its purpose "was to examine the philosophy of impact assessments in general and their desirability for and applicability to the Canadian situation. The intent of the workshop was to involve every participant in the formulation of answers to specific questions involving environmental impact assessments." Some of the questions and answers are relevant to offshore drilling projects in the Arctic since they come close to stating fundamental principles regarding EIA processes.

The participants of one working group agreed that "impact assessments are indeed the most favourable method at present to induce preservation of the environment." They concluded that EIAs "enable considerations to receive proper weight with other considerations in the evaluation of proposed actions in the public and private sectors. Published EIAs contribute to public understanding and acceptance of policies and decisions, to proper evaluation of aesthetic and other intangible values, to the development of baseline research and to the evolution of rules and standards for future actions." They concluded that "In principle a hearing is a necessary part of an environmental impact procedure unless the proposer of the action can demonstrate . . . that the formal hearing is not needed in the particular circumstances." To the question, Who should be required to prepare environmental impact assessments? the group replied "The proposer of any action having significant environmental impact should prepare the assessment. The adversary system (onus of proof on proposer) should be employed to ensure that there is full evaluation of the assessment. For this purpose government agencies should have the staff and funds necessary to make competent evaluations. Public interest groups should play an important role in this evaluation."

The group concluded that the general text of an EIA should be written primarily for the public and the decision-making body, and the research reports and data primarily for scientific analysis. They concluded, "The assessment should be distributed to the intervenors and be readily available to the public."

Another working group was asked, How can adequate scientific evaluation of information occur when much governmental and private data are proprietary? Even though two-thirds of the members of the group were from government and industry, the question was answered unequivocally: "Information pertaining only to the environment should not be proprietary. All research funded by public money, or carried out on crown land, should be publicly released, so should all environmental data from private enterprises after a project permit application has been filed. Adequate time should be allowed to use these data before the project decision is made."

Environmental Assessment in the Canadian Arctic

The application of these basic principles to environmental impact assessment in the North is inconsistent. Procedures related to consideration of the Mackenzie Valley pipeline came very close to honouring them.[31] The baseline data and the EIA prepared by Canadian Arctic Gas were publicly available nine months before public hearings began; they were evaluated by a government assessment group and its report is also a public document. Government agencies conducted a separate research programme and one consortium established an independent board to examine the potential effects on the environment of a natural gas pipeline. Finally, the government established the Berger Commission to hold thorough public hearings on the pipeline.

On the other hand, industry and government approaches to exploratory drilling in the Beaufort Sea and in other areas of the Arctic are less consistent with the basic principles of assessment. In the cases of drilling in the Arctic Islands and Hudson Bay, and the proposal to drill in Lancaster Sound, virtually every principle enunciated in the EPB workshop has been violated. The research programmes have been superficial, and the EIAs that were obtained were highly biased in favour of the proponent and both are held confidential under proprietary interest regulations. Public hearings were not held nor are they being considered.

In almost every chapter of this book there is reference to some aspect of the relationship between DINA and DOE on environmental matters. In Chapters 2 and 3 there is a detailed review of the DINA-DOE interface on offshore drilling in the Beaufort Sea. These accounts show that DOE as the agent for environmental protection has been ineffective in dealing with DINA, the determined advocate of northern development of the 1960 vintage.

DOE internal documents are convincing evidence that a feeling for environmental protection has been either absent or muted at senior levels of the department. Discussions with civil servants, and a recent review[32] of how federal policies are formulated and implemented, indicate that the pressures for northern development are still very strong in senior government committees such as the Advisory Committee on Northern Development, the Task Force on Northern Oil Development, and in Cabinet. In this climate it would require very strong determination on the part of the Minister of the Environment and senior officials to effect a reasonable balance between environmental protection and resource development in the North. Neither the past nor the present administrations of Environment Canada seems to have had that determination.

Ten days after Cabinet had approved the Offshore Drilling Memorandum, Jack Davis, then Minister of Environment, called his special assistant with some suggestions about a letter that the Deputy Minister was drafting to the Canadian Environmental Advisory Council. In a memo that lays bare the attitude of senior members of DOE, Mr. Davis' assistant passed his comments on to the Deputy Minister:

Another point the Minister raised in his telephone response to this letter was that you bear in mind the recent Cabinet approval of the Oil Drilling in the Beaufort Sea paper. This approval for the granting of permits to drill, has been given without enough time to do proper environmental assessment work. If the Council knew that we were about to go ahead with this work and had their environmental impact model handy, we could end up caught in the glue! (Appendix 2)

The Environmental Assessment and Review Process (EARP) announced late in 1974 is indicative of the department's weak approach.[33] The assessment procedures leave much to the developer's initiative, even the decision on whether or not the development will have an impact requiring assessment:

In this whole process, the proponent is bound by little more than his own discretion. The only rule he must follow is an undefined "rule of reason". Even if he concludes that there may be environmental impact associated with a project, a preliminary assessment requires no new research — only a review of existing information to determine whether the impact will be "minor" or "major". No further study is necessary unless the Screening Committee disagrees with the preliminary conclusion. Only when an action is classified as "major" is a formal impact statement required. This procedure will be invaluable for unearthing the obvious. Unfortunately, the purpose of environmental impact assessment is to create a tool whereby the obscure is identified.

By allowing the developer to make the primary decision on whether an assessment is necessary, the government has forfeited the benefits of impact assessment. By their nature, government agencies and private industry are development-oriented. The goal of impact assessment is to force an agency to look beyond the narrow horizons of economics and technology. But a policy that leaves the proponent the choice of when to consider the very questions he has failed to raise in the past is nonsensical. It is precisely the proponent's failure in the past that has given rise to the assessment concept.

When the government first outlined its assessment policy, the former Minister of Environment, Mr. Davis, told the Commons: "Public disclosure is important . . . . Public participation is vital." The detailed guidelines handed down in September, however, make it obvious that this is empty rhetoric and that the government does not have any intention of paying even lip service to the concept of public participation.

The public's opportunity to participate rests solely at the discretion of the Minister of Environment and the proponent. Only when they consider it "appropriate" will the Panel hold public hearings. There is no other provision to encourage or even tolerate the public. The rights of the private citizen are restricted to reading the preliminary assessments and formal statements submitted to the Screening Committee or the Panel. There is no right to comment on the documents, nor even the right to see the recommendations of the Panel. Even these extremely limited rights carry a proviso. If, in the judgement of the Minister or the proponent, a project may be jeopardized by premature disclosure, the public will receive no information at all.[34]

This extremely weak concept of environmental assessment for projects which come under federal jurisdiction is at the root of the weakness which exists in the assessment of Arctic offshore drilling. The Canadian assessment procedure is unlikely to do much to preserve the Arctic marine environment in the face of the intensive programme of offshore drilling which is underway and the weaknesses which exist in the implementation of the assessment process.

Environmental Assessment in the Alaskan Offshore

In contrast, the environmental assessment of offshore drilling operations is much more thorough in the United States. Also, each stage of the process is open to public participation and subject to review at public hearings.

The initial research programme for the Gulf of Alaska, the southeastern Bering Sea, and the Beaufort Sea will last 18 months. It will cost an estimated $18.4 million,[35] and the overall programme will require four to five years to complete. In a reversal of the Canadian situation, the basic assessment will be completed in advance of the sale of oil and gas leases anticipated late in the decade. As well, there will be additional assessment programmes related to specific projects as the development of leases is undertaken. The results of the advance research programme will be applied in these assessments, but additional investigations may also be made at that time if they are required.

There have been two other important national projects conducted as background to the leasing of new offshore regions, which in effect has been deferred since 1969 when the blowout of an oil well occurred near Santa Barbara, California. The first was *A technology assessment of Outer Continental Shelf oil and gas operations,*[36] a study funded by the National Science Foundation and by an interdisciplinary research team under the aegis of the Science and Public Policy Program at the University of Oklahoma.[37] The second was an environmental assessment of offshore oil and gas operations produced by the Council on Environmental Quality as a result of a request by the President in 1973. It covered the Atlantic and the Gulf of Alaska, and was based on existing knowledge. It identified knowledge gaps and made recommendations such as this one: "Carefully designed environmental studies should be initiated immediately in potential leasing areas and should be an essential and continuing part of OCS management. Such studies should be closely monitored and coordinated so that information can be integrated into ongoing operations and the results applied to decisions on leasing and regulating new areas. Special attention should be focused on determining long-term or synergistic effects of oil and other pollutants, if any, on marine organisms."[38]

This coordinated approach to the environmental assessment of offshore exploration has developed as a result of the National Environmental Policy Act which was passed by the U.S. Congress in 1970.[39] The act resulted in the formation of CEQ and institutionalized the use of environmental impact procedures when the federal Government is involved either as the proponent or as the licensing agent. The legal actions which stalled the trans-Alaskan oil pipeline for over three years were initiated by public-interest groups to enforce this legislation.

The courts play an important role in the American process — a role that could not easily be transferred to the Canadian structure. At the very least, however, a screening board to determine what actions should be assessed ought to be established under Canadian law. The decisions of the screening board could be enforced by court action.

There is little doubt that the lessons in the U.S. approach to environmental assessment would have been valuable in establishing a better balanced approach to offshore drilling in the Arctic. The environmental assessment and review process instituted by DOE is a token effort compared to the U.S. procedure which can truly influence decisions and activities related to the development of offshore petroleum resources.

Research for Environmental Assessment and its Application

Since the start of this decade many of the scientists who best know the Canadian Arctic have pointed to the abysmal lack of scientific knowledge of the Arctic marine environment. That lack of knowledge resulted in the crash programme of research for the Beaufort Sea. The same scientists would probably all state unequivocally that predictions on the possible effects of an oil spill cannot be based on observational and survey data collected in the time available to the Beaufort Sea project. Nonetheless, individuals from some well-known oil companies have been telling native people in the Arctic that the project has proven that they do not have to worry about the effect of oil on seals. The research which "proved" this involved six seals which were placed in tanks where they swam in water, covered with a crude oil film for 24 hours.[40] Studies of the animals following the experiment suggested that such an exposure does not cause permanent damage to healthy animals. Other seals tested in a laboratory, however, all died within 70 minutes. As one of the scientists stated in a letter:[41]

The questions still to be answered are: What are the effects of long-term exposure; the effects of exposure on seals subjected to the stress of unusual oil conditions and diminished food sources? What are the effects on new born pups in birth lairs or on seals in moulting condition? There are many open questions in other areas of the study, both physical and biological, including the matter of oil clean-up. All of this leaves me with an uneasy feeling about the state of our present knowledge, should a blowout occur.

And in virtually every area of the Beaufort Sea project there are as many or more open questions, because our basic knowledge of Arctic marine ecosystems is inadequate. The void should be filled to meet the needs of environmental protection in this century. We need a major programme of basic research in the Canadian Arctic. Canadians have a right to ask what Government intends to do to fulfill this basic need.

In addition to the need for a carefully developed long-term programme of basic research, a more specialized environmental impact research programme is needed. The highest priority would be research designed to identify or predict problems which could result from mining, petroleum, transportation, and water resource projects. The research should involve environmental studies and monitoring of ongoing projects. The objectives of this programme would be to determine what can be done to adapt exploration and production activities and methods in order to reduce the threat of environmental damage now posed by industrial development.

The third area for research is the application of environmental impact assessment to major projects, or to regions such as Mackenzie Bay where cumulative effects could result from industrial activity. In view of the inadequacies of DOE's environmental assessment review programme, what can be done to improve the present situation? One approach would be to shake hell out of DOE's environmental assessment review programme, forcing it to evolve into an effective process. This could be done by subjecting every assessment it makes (for example, the Mackenzie Delta Gas Producers application; or deepwater drilling in the Beaufort Sea) to an intense, critical review, and widely publicizing the results of each analysis.

But who could, or would, do the job? Our scientists and engineers could but it is doubtful they would; they are a tame lot in Canada, and have done little in public that Bay Street[42] would frown at.

In retrospect, the Treasury Board's refusal to pay 50% of the cost of the Beaufort Sea research was irrational, considering the Government's failure to conduct, or require industry to conduct, an adequate programme of basic research during the 1960s. In future, it would make more sense for the Government to pay for the basic research programme and for industry to pay for, but not control, research, surveys, and other activities associated with environmental assessment. As pointed out earlier, industry's record in this respect is much better than government's. Canadians can protect the Arctic environment if they really decide to do it.

References and Notes

1 Yates, A.B., *Energy and Canada's North: The search for oil and gas, Nature Canada,* July/September 1972.

2 Based on list of APOA research projects provided by DINA.

3 DINA, Oil and Minerals Division, *A position on oil and gas exploratory drilling in the offshore regions of Canada's Arctic* (Ottawa: DINA, Northern Natural Resources Branch, April 1973).

4 Chrétien, J., *Oil and gas exploratory drilling offshore, Northern Canada,* Draft Memorandum to Cabinet, May 1973, see Appendix 1.

5 This history of DOE involvement is traced in Chapter 2.

6 The list of proposed studies was considered by AWOGAC at its meeting on 5 October 1973. Our knowledge of this sequence of events was based mainly on personal discussions with members of DINA and DOE.

7 APOA news release, *$2 million for Beaufort Sea study,* 18 February 1974.

8 The statement undoubtedly referred to the Committee for the Original Peoples' Entitlement (COPE) which had put out its news release on offshore drilling on 8 February 1974.

9 DINA news release, *Beaufort Sea — No drilling until 1976,* 6 March 1974.

10 Milne, A.R., *Beaufort Sea and Mackenzie River Delta environmental studies,* talk given at CSPG-GAC International Symposium, Calgary, October 1974.

11 Woodward, H.W., *Other departments interests and channels for government industry cooperation,* Northern Canada Offshore Drilling Meeting, Ottawa, 1972.

12 Slaney, F.F., *Environmental field programme, Taglu-Richards Island, Mackenzie Delta,* Imperial Oil, January 1973.

13 Letter from R.E. Webb, Chief Ecologist, Slaney; included with report cited in n. 12.

14 Slaney, F.F., *Environmental impact assessment, Immerk artificial island construction Mackenzie Bay, N.W.T.*, Imperial Oil, 1972.

15 Woodford, J., *The violated vision* (Toronto: McClelland and Stewart, 1972).

16 Letter from R.E. Webb, Chief Ecologist, Slaney, included with report cited in n. 14.

17 Slaney, F.F., *White whale study in the Herschel Island — Cape Dalhousie coastal region of the Beaufort Sea*, Imperial Oil, 1974.

18 Imperial Oil's experience with the prediction of weather conditions from hindcasts during the construction of Immerk illustrates the need for a critical examination of commissioned reports. An article which described the construction of the island noted that "weather conditions were more severe than predicted from hindcasts. Strong winds and rough seas were commonplace. During the month of September only nine working days were realized and on September 22, icing conditions dictated that operations be suspended for the year. Of the 70 potential working days from July 15 to September 22, actual placement of material on the island occupied only 27 days." Riley, J.G., *How Imperial built first Arctic island, Petroleum Engineer*, January 1974.

19 The Beaufort Sea project is described in some detail in papers by J. Hnatiuk, *An environmental research programme for drilling in the Canadian Beaufort Sea*, paper presented at the 26th Technical Meeting of the Petroleum Society of the CIM, June 1975, and Milne, see n. 10. They are the industry and government managers of the project.

20 DINA, *Beaufort Sea project*, Bulletin No. 9, 6 November 1974.

21 Milne, see n. 10.

22 See Chapter 4 for insight into potential problems of oil spills from offshore production and transportation facilities.

23 On this point, the Minister of Indian and Northern Affairs gave approval for the construction of the drilling systems in March 1974 and drillships will be on their way to the Beaufort Sea before the validity of environmental data can be appraised.

24 Kendel, R.E., et al., *Distribution, population and food habits of fish in the western coastal Beaufort Sea*, Beaufort Sea Interim Report B1, DINA, December 1974.

25 Grainger, E.H., *Biological productivity of the Southern Beaufort Sea, The physical chemical environment and the plankton*, Beaufort Sea Interim Report B6 (part 1), DINA, December 1974.

26 Pimlott, D.H., *Offshore drilling in the Beaufort Sea*, COPE, January 1974.

27 The paper by Greene and MacKay, Appendix 6, shows how little we know about the behaviour of oil, and methods of containing and cleaning it up, in Arctic marine systems.

28 The study conducted in April 1975 of the behaviour of oil in an ice covered area involved two discharges, each of 180 gallons (approximately 5 barrels); by comparison, a scenario for a blowout of an oil well in the Beaufort Sea suggested that an initial discharge rate of 2500 barrels per day would be reduced to 1000 barrels per day by the end of a month. It was postulated that a flow of this magnitude would continue until the hole was plugged.

29 The tendency of industry representatives to over-generalize from preliminary data is illustrated in the paper by J. Hnatiuk, see n. 19.

30 Morley, G., and Odlum, B., editors, *Proceedings*, National Conference on Environmental Impact Assessment: Philosophy and Methodology (Winnipeg: Agassiz Centre for Water Studies, 1973).

31 *Northern Perspectives, The Mackenzie Valley Pipeline Inquiry*, 1974, and *We are embarked on a consideration of the future of a great river valley and its people*, 1975.

32 Dosman, E.G., *The national interest.* (Toronto: McClelland and Stewart, 1975).

33 *Northern Perspectives, The impact policy — empty rhetoric,* 1974 and *Federal environmental assessment policy,* 1975.

34 *Northern Perspectives,* see n. 33.

35 U.S. Department of Commerce, National Oceanic and Atmospheric Administration (NOAA), *Programme proposal: Environmental assessment of the Alaska Continental Shelf* (Washington, D.C.,: U.S. Printing Bureau, April 1975).

36 In the foreword to the book (see n. 37), technology assessment is described as "a class of policy studies intended to anticipate and explore the full range of consequences of the introduction of new technology or the expansion of oil technology in new and different ways." In describing what is new about this kind of assessment, it is brought out that one of its features "is its focus as a policy study on informing the interested public and the decision-makers of the possible ranges of consequences for new actions."

37 Kash, D.E., and White, I.L., *Energy under the oceans: A technology assessment of outer continental shelf oil and gas operations.* (Norman, Oklahoma: University of Oklahoma Press, 1973).

38 Council on Environmental Quality, *OCS oil and gas — an environmental assessment,* Vol. 1 (Washington, D.C.: Government Printing Office, 1974).

39 The history of the NEPA and a discussion of aspects of the experience gained in applying it are described in two papers in the proceedings of the EPB workshop (see n. 30). They are both authoritative discussions since the one on the Act was written by a former counsel to CEQ (W.T. Lake) and the second was written by the senior environmental project manager (R.A. Purple) of the U.S. Atomic Energy Commission. A thorough discussion of the Act and its impact is found in *Third annual report of the Council on Environmental Quality,* "NEPA — reform in government decision making" (Washington, D.C.,: Government Printing Office, 1972).

40 Smith, T.G., and Geraci, J.R., *The effect of contact and ingestion of crude oil on ringed seals of the Beaufort Sea,* Interim Report of Beaufort Sea Project study A5, December 1974.

41 Letter from J. Geraci, Associate Professor, Wildlife Diseases, to K. Sam, NAG, 14 March

42 For the information of U.S. readers, Bay Street is to Canada what Wall Street is to the States.

Chapter 10

Oil Under the Ice — In Retrospect

Offshore exploration for oil and gas is a new dimension in the search for Arctic petroleum reserves. It entails far higher risks to the environment and to the interests of native populations than land-based operations. The failure of the federal Government to assess and protect these interests is not an isolated failure; it is part of a much broader framework of government actions encouraging rapid resource exploitation at the expense of social and environmental goals, and the exact reverse of the Government's official northern policy announced in 1972. The 1972 policy put the concerns of native people first, environmental protection second, and industrial development third. But, as events in the offshore reveal, this ranking of priorities is mere sleight-of-hand. In practice, the imperatives of the oil industry have determined the course of events.[1]

Able to state its case forcefully and having easy access to the highest Government officials, the industry has had little difficulty in dictating the pace of offshore development. By the time the companies were ready to venture onto the Arctic Ocean, the industry had already established within government a powerful, perhaps irresistible momentum for resource development. Still, it was clear that the Government had no coherent vision of how offshore development could proceed in the public interest. In contrast to countries such as Norway and the U.S. where offshore exploration has been carefully planned and tightly controlled, Canada's northern affairs bureaucracy was prepared simply to respond to industry initiatives and act as a mirror to industry interests. The memorandum to Cabinet on offshore drilling submitted by DINA made this clear. Although offshore exploration permits had been granted haphazardly for up to a dozen years beforehand, the department had given no serious thought to how and when development of the Beaufort Sea should proceed. In the absence of any planning on its own part, the easiest course was simply to adopt the industry position as its own.

In the High Arctic, it was even harder to separate the industry's interest from the Government's. In September 1973, a Panarctic vice-president wrote to the head of DINA's Oil and Minerals Division pushing for approval of Panarctic's first offshore well: "it is extremely important to Panarctic and its participants to develop the means of offshore drilling at the earliest possible date. If offshore drilling is delayed and Panarctic does not make discoveries in its Paleozoic well, the assembling of the required threshold reserves will be delayed. The Hecla 0-62 well is essential if we are to be allowed to make a prudent lease selection." The recipient of the letter must have wondered what side of the fence he was on, since his immediate superior was a member of Panarctic's Board of Directors.

DINA has always been acutely aware of protecting its hegemony in the North. Although the department has little environmental expertise, it has consistently excluded the Department of the Environment from a direct role in the decision-making process; environmental considerations have never been more than afterthoughts in processing drilling applications. This strategy has not always worked to DINA's best advantage. The Beaufort Sea environmental information provided to Cabinet, for example, was embarrassingly inadequate and the department was subsequently forced to retreat from its hard-line position. Nevertheless it succeeded in excluding DOE from exercising an effective regulatory role on environmental aspects of offshore drilling, just as it had on land-based operations.

DINA's control of environmental protection in the Northwest Territories and the Yukon is one of the central issues raised by offshore drilling in Arctic waters.

Part of the reason for DINA's failure to maintain any control over the shape and pace of events was its inability, or perhaps unwillingness, to comprehend broader social and environmental interests. Having firmly allied itself with industry, environmental and native concerns came to be seen as complications rather than as integral parts of the public interest. DINA adopted a course that has come to typify northern decision-making: it kept everything secret. The easiest way to deal with natives and environmentalists was simply not to deal with them at all. But it would not have been possible to use such a stratagem if information on development proposals had been available to Parliament, and through it, to the public at large. The use of secrecy to give resource development *de facto* priority over social and environmental goals, is one of the important issues which must be resolved if development is to be brought into perspective.

Shortly after the release of the COPE report on offshore drilling, and months after Cabinet had approved drilling in the Beaufort Sea, DINA's Assistant Deputy Minister for Northern Affairs defended the department's failure to consult with western Arctic communities by saying that there was really nothing to consult them about. What is frightening in such a statement is not that it was made, but that the most senior bureaucrat for Northern Affairs probably believed it. This kind of insensitivity to the interests of northern native people has been, in many other instances in the past, a distinguishing feature of northern policy. Much of the problem again lies in DINA's espousal of industry's world view, and in that view there is no room for such eccentricities as the social and cultural values of renewable resources. Native people, to the extent that they are part of the industry picture at all,are simply potential job-holders. It would be unrealistic, perhaps, to expect the extractive industry to have any other concept of the North, but when DINA takes the same view, it is not hard to understand the disillusionment of northern natives. Ironically, it is precisely by excluding native interests and concerns that DINA forfeited its only opportunity to initiate creative, humane policies in the North. After all, where is the challenge in being the handmaiden of industry?

Oil companies operating in the North are fond of talking, particularly in the South, about the great risks that they are taking. Their public relations campaigns point out that the companies are risking millions of dollars on the search for oil which they may never find. And so they are. But it is an informed risk; no one is forcing the companies to drill. Dome Petroleum has no doubt considered carefully the financial risks and the financial rewards of drilling in the Beaufort Sea, and has done so on the basis of the best and most reliable geological and engineering data available.

In contrast, consider the problem the native population in the western Arctic faces. They are expected to bear the social and environmental costs of development, even when they dispute the very right of developers to be there before their land claims are settled. In the meantime, native people must accept the possibility of damage to the integrity of the lands and waters they traditionally use. At stake are values that are important not only to individuals but to the Inuit society. Provided with information designed to persuade rather than inform, they are right to be anxious about their future. They will pay for the mistakes.

This failure to consider the interests, needs, and rights of native people while promoting petroleum development is another of the central issues raised by offshore drilling in the Arctic. There is an opportunity to redress it in the settlement of land claims. Whether the opportunity will be taken or not remains to be seen.

In the course of our investigation we were told time and time again of the determination of oil companies to protect the environment. Can the petroleum industry, in fact, be trusted to establish high environmental standards in its rush to discover and develop the oil and gas resources of the Arctic? W.O. Twaits, Chairman of Imperial Oil, stated the industry case in unequivocal terms: "We regard it as one of the most important matters facing the company. Protection of the environment is not a passing fad. Pollution will not go away by itself. Indeed, as population and economic growth increase so will the potential for despoiling the land, air and water. This means that every person within Imperial Oil has a responsibility in helping combat pollution of any kind. Preserving the environment is no longer just a good public-spirited thing to do. It is a must."[2]

But our investigation of offshore drilling indicates that the industry concept of adequate environmental protection differs significantly from the one in which conservationists and native people believe. The general inaccessibility of environmental reports and impact assessments does not allow opportunity for critical review of the ones which are conducted by or for the industry. Two specific aspects of Imperial's approach to environmental assessment highlight the problem as seen from a conservationist's viewpoint: the company has placed restrictive terms of reference on consultants' reports and has avoided conducting any studies of the impact of oil spills on the critical near-shore environment of the Beaufort Sea where their exploration programme is underway. It seems apparent that the company has too much at stake economically to go much beyond cosmetic environmental programmes of its own volition. It is apparent there must also be strong environmental regulations and rigid enforcement of them at every stage in the exploration and production of oil and gas. The approach of the industry to development of environmental units is of interest to environmental protection. In the majority of cases this has been accomplished by transferring professional staff, commonly engineers or petroleum geologists, who have been involved in exploration, development, or production aspects of the industry. Rarely do their environmental organizations include people who have not been steeped in the development/progress ethos of the industry. It seems unlikely that retreading production men as environmentalists will result in profound changes in the environmental perceptions or approaches of large industries involved in high capital risk projects in remote areas of the Arctic.

In retrospect, events in the history of offshore exploration make us believe a sense of perspective on the use of Arctic regions should surely be based on a sense of humility which our species has seldom achieved. But we must somehow find the proper humility if we are to change the frontier approaches which we seem compelled to use in commandeering resources for our use. A new perspective could logically arise from an appraisal of the needs and wants of mankind and of our relationship with all other living things. In the past we have found it difficult to learn from such an appraisal. Could it be that things have changed enough in the last decade to make it possible? If our concerns have real meaning perhaps we can gain the perspective which will allow us to utilize the resources of the Arctic without degrading it, without significantly affecting all the other species which make up these austere, beautiful, and vitally important world ecosystems.

References and Notes

1. *The National Interest* by E. Dosman (Toronto: McClelland and Stewart, 1975) provides detailed insights of the processes involved in the formulation of northern policy from 1968 to 1972. It deals with the roles played by deputy ministers and their peers and the interdepartmental committees on which they serve. It complements this study since our investigation of offshore drilling focussed at secondary and tertiary levels, always a full tier below that represented by Dosman. Assistant deputy ministers were at the lower edge of his vision while they were at the upper edge of ours. More squarely in our focus were directors, regional managers, and resource management officers. These are the men who implement policy directives and who enforce the regulations by which the environmental impact of offshore drilling is supposed to be maintained at tolerable levels.

2. Twaits, W. O., *The environment: whose responsibility? Imperial Oil Review,* vol. 56, no. 5, 1972.

Appendix 1

This Document is the Property of the Government of Canada*

Memorandum to Cabinet **Confidential**

Oil and Gas Exploratory Drilling Offshore, Northern Canada May 1973

Summary
Numerous holders of oil and gas permits issued by the Government of Canada covering Canada's northern continental margin, seek approval in principle from the Department of Indian Affairs and Northern Development for their proposals for the design, development, and construction of multi-million dollar drilling systems appropriate for exploration of Canada's Arctic offshore oil and gas resources.

Within the limits of present day oil and gas offshore technology and operating experience, the offshore drilling systems will incorporate the highest levels of fail-safe reliability available.

It is concluded in this Memorandum that federal legislative and administrative controls implemented by professionally qualified experts of departmental staffs (principally of the Department of Indian Affairs and Northern Development, Department of Transport, and Department of Environment) are adequate to ensure reasonable and effective safeguards and contingency plans for the protection of Canada's Arctic environment. Although the inherent risks of accidental pollution of the Arctic marine environment cannot be reduced to zero, the risks are considered to be low and reasonable with respect to probable national and regional economic and social benefits to be derived.

In view of the circumstances, it is recommended that the Government approve in principle the exploration for and development of the potential oil and gas resources in the offshore regions of Canada's Arctic by reconfirming the authority of the undersigned to approve and license such activities.

1. Problem:
The Government of Canada (through the Department of Indian Affairs and Northern Development and its predecessor department) has issued oil and gas exploratory permits for offshore regions in Canada's Arctic, dating as far back as 1960. The holders of certain of these permits have proposed exploratory drilling programs to the Department and are awaiting a decision. Exploratory drilling for oil and gas in the Arctic offshore holds promise of significant economic and social benefits on the one hand, but on the other hand, includes an inherent risk of accidental pollution. Within the limits of present day oil and gas offshore technology and operating experience, the offshore drilling systems proposed for the Arctic will incorporate the highest level of fail-safe reliability available. The risk of accidental pollution will be reduced to what may be considered to be an acceptable minimum, however, the minimal risk exists, and the public is aware of this.

*This document's content has not been altered by CARC, except to delete some irrelevant copy (ed.).

2. Objective:
To have Cabinet reconfirm the authority vested in the undersigned to approve and to licence the activities of the private sector for the exploration and ultimate development and production of the oil and gas resources of the offshore regions of Canada's Arctic.

3. Factors:
3.1 **Disposition of Canada's Oil and Gas Resources**
More than half of Canada's ultimate recoverable potential oil and gas is to be discovered in the North. The basins with highest potential in decreasing order are, the Mackenzie Delta-Beaufort Sea basins, the Arctic Island and contiguous sea basins, and the basins in the seabed of the Baffin Bay-Labrador Shelf. The greater part of the potential of northern Canada is offshore and covered by sea-ice for most of the year. Special technologies and relatively high costs are required to explore for and ultimately to develop and produce these resources . . . .

3.2 **Present Status — Permits Issued and Exploration Activities**
To date, permits have been issued for over 440 million acres in northern Canada, essentially all of the prospective sedimentary areas, including the offshore regions to water depths of 1500 feet. In January 1969, a sale by tender of five blocks of permits, in water depths of up to 600 feet and ice-covered with the exception of some eight to ten weeks in the summer, in the Beaufort Sea . . . brought bids in the form of commitments to spend over $15 million on exploratory work (work bonus) within the six-year period ending 30 January 1975. Exploration expenditures, in accordance with the work performance obligations of the permits, have climbed to some $230 million in 1972. The exploration expenditures with respect to the offshore permit areas have been principally for seismic surveys and research related to exploratory drilling.

3.3 **Discoveries on Canada's Northern Coasts and Seabed Extension**
Over the past four years Panarctic Oils Ltd. has discovered gas on the west and east sides of the Sabine Peninsula of Melville Island and on King Christian, Thor and Ellef Ringnes Islands, and has found oil in wells on Ellef Ringnes and Ellesmere Islands. Several of these gas structures extend into the seabed. In the Mackenzie Delta region and coastal tracts of the Beaufort Sea, Imperial Oil Ltd. has discovered oil at Atkinson, Mayogiak, and Atertak, and gas at Ivik, Mallik, and Taglu. Gulf Oil Canada Ltd. has found gas at Parsons Lake, Titalik, Ya-Ya, and Reindeer, and Shell has discovered gas at Niglintgak . . . . These discoveries together with seismic surveys of the delta, coastal tracts, and adjacent seabed of the Beaufort Sea point to excellent prospects for oil and gas in the offshore.

3.4 **Industry's Proposals to Drill Offshore, Northern Canada**
3.4.1 Design concepts have been developed for offshore drilling systems, which appear appropriate for drilling in the Arctic marine environment, from the detailed information obtained from studies sponsored by the Arctic Petroleum Operators Association (APOA) and from information derived from previous design, construction, and the operating experience.

3.4.2 In 1972 the Department of Indian Affairs and Northern Development received proposals for drilling systems for operating in the open-water season of the southern Beaufort Sea, essentially in the area defined as Zone 12 in the Shipping Safety Control Zone Order pursuant to the Arctic Waters Pollution Prevention Act from the Beaufort Sea Task Force (BSTF), and Hunt International Petroleum Company (HIPCO). The BSTF represents jointly: Amoco Canada Petroleum Company Ltd., Aquitaine Company of Canada Ltd., Canadian Superior Oil Ltd., Elf Oil Exploration and Production Canada Ltd., Gulf Oil Canada Ltd., Hudson's Bay Oil and Gas Company Ltd., Mobil Oil Canada Ltd., Texaco Exploration

Canada Ltd., and Union Oil Company of Canada Ltd., who variously hold five blocks of the work bonus permits. The HIPCO and the BSTF proposals are basically similar, each consisting of a barge-type drilling vessel with a centrally mounted conventional drilling rig, supported by two workboats and helicopters from a supply base . . . .

3.4.3 The proposed drilling systems will each cost in the order of $30 million to design, construct, and commission and will cost about $75,000 per day when operating and about $30,000 per day including fixed costs and amortization, while shut down for the winter season. All of the exploratory wells drilling during the proposed programs will be abandoned.

3.4.4 Before proceeding with the large investments for the design and construction, the sponsors have sought approval in principle from the Department for their proposals to ensure that when completed the systems will be authorized to drill exploratory wells on Canada oil and gas permits in Arctic waters.

3.5 **Legislation and Control of Drilling Offshore, Northern Canada**
Requirements for drilling systems operating in the southern Beaufort Sea during the open-water season, attached to a draft letter of approval in principle , . . . are developed from inter-departmental consultation and consideration of requirements under the Oil and Gas Production and Conservation Act, the Arctic Waters Pollution Prevention Act and Regulations, the Territorial Land Use Regulations, among other items and include requirements concerning financial responsibility and Canadian content.

3.6 **Risk of Damage to the Environment Offshore, Northern Canada**
3.6.1 The major environmental concern with oil and gas operations in the offshore regions of Canada's Arctic is and will likely continue to be that of a major oil spill such as could occur from an oil well blowout. It is difficult to quantify the consequences of the release of oil into Arctic waters; however, the timing and extent of a spill, meteorological conditions, the occurrence of ice, the character of the oil, and the low temperature environment would be among the most significant factors.

3.6.2 A blowout, that is the uncontrolled flow-back of drilling mud from a well that enables oil, gas, or water to flow to the surface, can only happen when the pressure of fluids contained within a porous formation exposed to the wellbore exceeds that of the hydrostatic head of the mud column and requires the simultaneous failure of all wellhead controls systems. In the narrowest sense, blowouts are a result of human failure that manifests itself in the form of either improper selection, installation, maintenance, or testing of equipment, or improper human action.

3.6.3 The record of offshore oil and gas well blowouts is not without merit. Out of over 20,000 offshore wells drilled throughout the world, but particularly off the coasts of United States, to the beginning of 1972, there had been 47 blowouts reported. Only four out of the 47 blowouts caused major oil spills, two in the Gulf of Mexico, one in California, and one in the North Sea. Forty-one of the blowouts were gas and 30 of the gas well blowouts bridged themselves over and stopped blowing unassisted, as commonly happens when gas wells blow out. Relief wells were required to be drilled from offsetting locations to control nine of the blowouts. Offshore oil and gas development precedents in the ice-infested waters of Cook Inlet, Alaska, the cold northern waters of the North Sea, and the cold southern waters of the Bass Strait of Australia illustrate that discoveries can be developed without apparent damage to related marine ecosystems.

3.6.4 The Arctic offshore drilling rigs will incorporate the highest level of fail-safe reliability within the limits of present day offshore technology. Operators will be required to submit a Waste Management Program outlining methods proposed for disposing of polluting and toxic wastes generated during the course of the drilling operations and a detailed Blowout Control Contingency Plan outlining provisions to prevent and control blowouts, and a detailed Oil Spill Contingency Plan outlining provisions to control, contain, and clean up oil spills.

3.6.5 Floating offshore drilling rigs, such as the barge rigs proposed for the Arctic, follow fully developed and documented shut-in and quick release procedures in the event of an emergency such as a severe storm, or invading iceflows. The Arctic offshore rigs will be additionally equipped to be rapidly moved off location in the unlikely event of a complete loss of well control resulting in a blowout by shearing anchor chains on one end of the vessel and winching forward on the anchors attached to the other end of the vessel, after releasing from the wellhead.

3.6.6 If it is necessary to drill a relief well to control an offshore blowout the primary offshore drilling rig will normally be relocated and restocked with backup equipment and supplies maintained in the area for such contingency and then proceed to drill the relief well. In the event the primary offshore drilling rig is inactivated as a result of a blowout, a second offshore drilling rig will be brought in to drill the relief well. The Blowout Control Contingency Plan to be submitted for approval will include a formal signed agreement with the owners and operators of the alternate offshore drilling rig or rigs that will be used to drill a relief well, if necessary, and a formal plan and schedule for drilling the relief well.

4. Alternatives:

Alternative 1

4.1 The Government may approve in principle the development of the potential oil and gas resources offshore northern Canada by confirming the authority of the undersigned to approve and to licence those proposals that will be considered to be satisfactorily designed, constructed, operated, and controlled to ensure minimal risks of environmental impairment and maximum benefits to Canada.

4.2 Government policy of previous years has been to stimulate the development of Canada's potential northern resources by issuing oil and gas permits with an obligation on the permittee to explore at an acceptable rate. The Government's Northern Policy for the 1970s imposes prior and due concern for the preservation of the environment and for social benefits, particularly for Northerners. Alternative 1 sets these potentially conflicting elements in reasonable balance and is consistent with the policy implicit in the Arctic Waters Shipping Pollution Regulations, Land Use Regulations, Northern Inland Waters Regulations and others.

Alternative 2

4.3 The Government may deny the development of northern offshore oil and gas resources for a specified period of years or for all time.

4.4 Alternative 2 is inconsistent with previous Government policy of providing a balanced program of industrial development and environmental protection with the consequent economic and social benefits to Northerners . . . .

5. Financial Considerations:

5.1 **Near Term**

5.1.1 During the offshore exploration phase the Government will be called upon to mount extensive Arctic oceanographic and other scientific investigations and researches to extend the present information base and to effect the Federal Contingency Plan.

5.1.2 Reliable meteorological and ice reconnaissance services in the operating region will increase the safety and efficiency of operations. The operators recognize this and have outlined provisions for surveillance in their local operating areas. The Atmospheric Environment Service (AES) of the Department of the Environment recognize this and have developed specifications for meteorological and ice services in the region.

5.2 **Longer Term**
5.2.1 The expectations of exploratory drilling in the offshore regions of northern Canada are to ultimately develop oil and gas reserves. The associated lease rentals and production royalties will result in substantial direct payments to the Government.

5.2.2 When sufficient offshore oil and gas reserves are proven to warrant development and the construction of a transportation system, the Government will be required to develop the required infrastructure.

5.2.3 A return on the Government's investments in Panarctic Oils Ltd. may hinge upon means to drill and produce the offshore extensions of its current and future gas discoveries such as Hecla, Drake Point, Thor, and King Christian.

5.2.4 Indirect revenues from taxation, reduction of future international payments for oil and gas supplies, and developments related to an Arctic oil and gas industry may be critical to the nation's future economic development.

5.2.5 The net benefits to the Government of Canada ultimately resulting from Arctic oil and gas operations are expected to be three or more times the comparable benefits which have and will accrue to Alberta from the developments of that province's conventional oil and gas resources.

6. Federal-Territorial Relations:
Drilling for and ultimately developing offshore oil and gas resources are of less direct concern to Territorial Governments than current onshore drilling, producing, transporting, refining, and marketing of oil and gas in the Yukon and the Northwest Territories. However, senior officials of the Territorial Governments have been kept appraised of industry proposals for drilling in the Arctic offshore and have committee representation to consider the environmental and social impact . . . .

7. Inter-departmental Consultation:
Department officials have consulted with colleagues in the Ministry of Transport with respect to drilling and service vessels and aircraft, and those in the Department of Environment with respect to environmental facets of the Arctic, possible pollution hazards, and relevant contingency planning. They have also consulted with colleagues in the Department of Energy, Mines and Resources with respect to their prior offshore drilling experience, particularly in Canada's east coast and in the Hudson Bay. Colleagues in the Department of Industry, Trade and Commerce have been consulted with regard to the means of ensuring the maximum Canadian content practical in the design, construction, equipping, and operation of the offshore drilling systems.

A two-day meeting, 5-6 December 1972, of over 180 senior officials of government and industry was arranged in Ottawa to discuss proposals for drilling in Canada's Arctic offshore.

8. Public Relations:
Either of the two alternatives will be subject to criticism, on the one hand from extremists in the environmental camp who wish to bank natural resources against future needs and from native organizations who pursue claims to a greater share in the resources and fear early development, and on the other hand from the busi-

ness sector who have either already made large investments or feel that development is needed to support the economy.

It will be necessary to keep the public informed of the decisions on these matters in the context of the evolution and implementation of the Government's Northern Policy, and to support that decision in the face of criticism.

9. Conclusions:

9.1 The established federal and administrative controls backed up by the professional expertise of departmental staffs are adequate to ensure reasonable and effective safeguards and contingency plans for the protection of Canada's Arctic marine and coastal environments with respect to industry's currently proposed and future activities in the Arctic offshore.

9.2 The proposals submitted by the Beaufort Sea Task Force Group, and by Hunt International Petroleum Company for offshore drilling systems to drill exploratory wells in the southern Beaufort Sea during the open-water season, may be approved in principle and the required Land Use Permits, Drilling Authorities, etc., may be issued where officials are satisfied that specifications and stipulations and other requirements have been, or will be, appropriately met.

9.3 Other proposals submitted for drilling in other areas offshore northern Canada, or to install facilities for producing oil and gas shall be deferred pending presentation to the Department of Indian and Northern Affairs of information and analyses adequate to assess the proposals particularly in respect to standards for safety, pollution prevention, and the maximum ultimate recovery of the oil and gas resources.

9.4 The risks of pollution and impairment of the Arctic environment cannot be reduced to zero, however, the risks are considered to be so low as to be acceptable for consideration of the national and regional benefits to be realized.

9.5 Industry and government agencies should be encouraged to coordinate planning, funding, and conducting studies of the Arctic marine environment having impact upon the design and development of new technologies to explore, discover, assess, inventory, and ultimately to develop the rich potential oil and gas resources in the seabed of Canada's northern continental margin for the benefit of Canada.

10. Recommendations:

It is recommended that the Government approve in principle the exploration for and development of the potential oil and gas resources in the offshore regions of Canada's Arctic by reconfirming the authority of the undersigned to approve and license such activities.

The undersigned may grant approval in principle to specific proposals to undertake activities such as exploratory drilling in the southern Beaufort Sea to assure and encourage industry to make the large pre-investments required for Arctic offshore oil and gas ventures, consistent with the Government's Northern Policy and subject to the operators meeting such stipulations and other reasonable conditions as may be identified.

Minister of Indian Affairs
and Northern Development

Table 1

ESTIMATED ULTIMATE RECOVERABLE POTENTIAL: OIL AND GAS FOR GEOGRAPHIC REGIONS OF CANADA

Region	Oil		Gas		Oil and gas combined	
	Billion bbls	% of total	Trillion cu. ft.	% of total	B.O.E.*	% of total
Cordilleran	1.1	1.29	10.9	1.89	2.9	1.6
Mackenzie — Banks Basins	8.0	9.39	64.0	11.08	18.7	10.3
Arctic Islands ex Arctic Lowlands	22.6	26.53	202.7	35.11	56.4	31.1
N.W.T. Mainland ex Mackenzie Delta	3.8	4.46	20.4	3.53	7.2	4.0
Western Canada Sedimentary Basin (including foothills)	20.1	23.59	124.2	21.51	41.0	22.5
Hudson Bay Plateau — Arctic Lowlands	6.0	7.04	6.0	1.04	7.0	3.9
Eastern Canada — Maritimes	1.5	1.76	16.6	2.87	4.0	2.2
Labrador Shelf — Baffin Bay	15.0	17.61	90.0	15.59	30.0	16.6
Scotian Shelf — Grand Banks	7.1	8.33	42.6	7.38	14.2	7.8

*B.O.E., Barrel of Oil Equivalent on basis 6000 mcf gas = 1 bbl. oil. Ratio of ultimate recoverable B.O.E. to proven B.O.E. = 6.9 to 1.

Table 2

TOTALS FOR CANADA

Sedimentary basin area (excluding continental slope)	Square miles
Onshore	1,412,000
Offshore	1,064,000
Total	2,476,000
Sedimentary volume (excluding continental slope)	3,538,000 cubic miles
Ultimate recoverable potential	
Oil	85 billion bbls.
Gas	577 trillion cubic feet
B.O.E.*	181 billion bbls.

*B.O.E., Barrel of Oil Equivalent on basis 6000 mcf gas = 1 bbl. oil. Ratio of ultimate recoverable B.O.E. to proven B.O.E. = 6.9 to 1.

Appendix 2

August 10th 1973

(Deputy Minister)

Letter to Chairman,
Cdn. Environmental Advisory Council*

While the Minister concurs generally with the text of the proposed letter to Dr. Porter, he would like you to bear in mind the fact that we need to put them to work (on one or specific projects).

Mr. Davis would like you to challenge them to pick a particular project for environmental impact assessment. He'd really like them to help us in our work in this area. In particular, the Minister would like to have one or two of the members of the Advisory Council on our Task Force on the Vancouver International Airport. The Advisory Council might be asked to produce a model for environmental impact.

Another point the Minister raised in his telephone response to this letter was that you bear in mind the recent Cabinet approval of the Oil Drilling in the Beaufort Sea paper. This approval for the granting of permits to drill has been given without enough time to do proper environmental assessment work. If the Council knew that we were about to go ahead with this work and had their environmental impact model handy, we could end up caught in the glue!

These comments might be useful to you in re-working your letter to Dr. Porter for the Minister's signature.

Thank you for your kind cooperation.

Special Assistant
Department of Environment

*This document's content has not been altered by CARC (ed.).

Appendix 3

Committee for Original Peoples Entitlement*
Post Office Box 1661
Inuvik, N.W.T.

Press Release 8 February 1974

Drilling for Oil and Gas in the Beaufort Sea

The Committee for Original Peoples Entitlement today expressed its concern over the prospect of offshore drilling for oil and gas in the Beaufort Sea. COPE's Board of Directors met at Paulatuk last weekend and directed that a request be made to the federal government for full consultation before offshore drilling is allowed to proceed. COPE considers the Beaufort Sea to be a high hazard area and is deeply concerned about the effects that oil spills or well blowouts could have on the marine environment and on the native settlements that border on, or are close to, the Sea.

Native people of this region have long been afraid of the possible environmental impact of offshore drilling but until lately have not been able to get much information about it. Recently, COPE asked Dr. D. H. Pimlott, University of Toronto zoologist presently working in the western Arctic as a resource person to native organizations, to prepare a report on the present status of offshore drilling proposals. Dr. Pimlott's work here is sponsored by the Canadian Arctic Resources Committee. The report, entitled "Offshore Drilling in the Beaufort Sea," documents the oil industries' plans for offshore drilling, the federal government's procedures for granting permission for such activity, and the current state of environmental knowledge of the impact of offshore drilling.

The report makes three outstanding points:

1. That offshore drilling represents an important new phase of northern oil exploration, involving unknown but possibly very significant risks to the environment. This environment, probably the most hazardous the oil industry has had to face anywhere in the world, could be very seriously damaged by accidental blowouts.

2. That this new phase of exploration, which has been approved in principle by the federal Cabinet, has been shrouded in secrecy. No local communities or native organizations in the North have been informed or consulted and no information has been provided to the Canadian public.

3. That the government's proposed program of environmental impact research is highly inadequate. In particular, only one year has been allowed for research, which is not enough. It appears the federal Cabinet has been misinformed about the potential hazards of offshore oil drilling, and that emergency procedures may not be adequate.

The Board of Directors of COPE is extremely disturbed by this report. It would appear that balanced, long-term development for the benefit of all northerners and indeed all Canadians is being sacrificed for immediate profit and a panic reaction to the energy crisis. COPE considers the risk of blowouts and oil spills in

*This document's content has not been altered by CARC (ed.).

the Arctic Ocean, especially on or under the sea ice, to be unacceptable at the present time. We have no evidence that there are adequate precautions for blowouts late in the season, or under conditions of heavy, drifting ice. There are several communities in this region which are directly or indirectly dependent on the sea. All around the edge of the Beaufort Sea the native people are trapping foxes or hunting white whales and seals and catching fish that run in from the Sea. In the winter the people from Holman Island, Paulatuk, Sachs Harbour, and Tuktoyaktuk travel far out onto the sea hunting polar bears. Some of the permit blocks where drilling will be done first are right in the area where the people from Tuktoyaktuk hunt polar bears. Others lie close to the coast in areas which are critical to seals, white whales, and waterfowl.

It is appalling that neither COPE, Inuit Tapirisat of Canada, Settlement Councils, nor hunters' and trappers' associations in the region have been consulted. COPE does not consider offshore drilling to be just another exploration operation to be approved when the time comes, as with land-based drilling. This is a new phase of exploration requiring the most careful consideration and precaution at the outset. COPE considers the record of the government and industry in taking adequate environmental precautions, in having adequate emergency procedures, and in enforcing existing environmental regulations, to be unsatisfactory. It is unacceptable that the Beaufort Sea, which is so important to the people of this region, to say nothing of the global population, should be subjected to such a risk, especially under the authority of an administration which has a proven record of consulting with native people and informing the Canadian public, after the fact.

COPE's Board of Directors has requested the federal government to begin immediate consultation with COPE and Inuit Tapirisat of Canada on the whole matter of offshore drilling in Arctic waters. The objective should be to develop an agreement which will include adequate safeguards for the interests of native people and their environment.

There is no desire to obstruct exploration unnecessarily but the vital interests of the native people must be protected. Verbal promises will not do. Sam Raddi, President of COPE, has already written to the Ministers of Indian Affairs and Northern Development and of Environment Canada asking them to delay permission for offshore drilling for at least three years, in order that proper studies can be conducted.

On behalf of the native people of the Northwest Territories, COPE asks for the support of the Canadian public in resolving this serious matter. While native people are in a position to be most directly affected, offshore drilling in this critical area warrants world attention and consideration. COPE particularly requests conservation, environmental, and scientific organizations to consider the potential implications of this new phase of exploration for oil and gas in the Arctic.

Appendix 4

Biology of the Beaufort Sea

Thomas G. Smith*

Introduction

The Beaufort Sea has been defined as the area bounded to the north by a line from Barrow to Prince Patrick Island and to the east by a line from Cape Kellet, southern Banks Island to Cape Parry (Wilson 1974). The area east of this, called the Amundsen Gulf, is strongly influenced by the physical conditions and fauna of the adjacent Beaufort Sea area.

Since the emphasis in this paper is on the biological resources of most direct importance to the Inuit populations, I include the Amundsen Gulf area in this discussion.

In the Canadian Arctic there are two villages immediately bordering the Beaufort Sea: Tuktoyaktuk on the mainland and Sachs Harbour on the west coast of Banks Island. Two other villages, Paulatuk in southern Darnley Bay and Holman on western Victoria Island, are situated in the Amundsen Gulf. A small number of Inuit hunters, from Inuvik and Aklavik, also make hunting trips along the mainland coast between Baillie Island and Herschel Island. The total Canadian Inuit population bordering the Beaufort Sea and Amundsen Gulf coasts is approximately 1075 (N.W.T. Community Data, 1972). The total Alaskan coastal Inuit population from four communities between Point Hope in the west and Barrow in the east is approximately 2900 (U.S. Bureau of the Census, 1970).

Biology

Oceanography

Little is known about the physical, chemical, and biological oceanography of the Alaskan North Slope and the Canadian southern Beaufort Sea. There appear to be two distinct biological habitats, one estuarine, the other marine, along the southern Beaufort Sea coast. The Mackenzie River with its high annual discharge (about 7.5 million litres per second) creates an area of low salinity and high turbidity between Cape Dalhousie to the east and Herschel Island in the west. Oxygen and chlorophyll data indicate a low inshore primary production probably due mainly to low light penetration because of the high river-induced turbidity (Grainger 1974). The same situation but to a lesser degree exists near the mouth of the Colville River on the Alaskan north slope (Alexander 1974).

In more truly marine areas, further removed from the influence of river discharge, a higher primary and secondary production and diversity is indicated (Grainger 1974, Wacasey 1974).

Birds

In the Beaufort Sea area most species are aquatic and migratory. Barry (1973) lists 24 species which use the west-to-east migration route, following and feeding

*Arctic Biological Station, Fisheries and Marine Service.

in the open leads to their summer nesting areas along the southern Arctic mainland coast, or turning north to the higher Arctic islands. A further 33 aquatic species are listed by Smith (1973a) for the Amundsen Gulf area adjacent to Holman and Victoria Island. This area includes a number of aquatic species usually found in the eastern Arctic.

In dry years on the southern prairies over 335,000 ducks have been estimated (Barry 1973) in the Mackenzie Delta region. Smith et al. (1964) estimate that 8000 geese and 4000 swans nest in the outer Mackenzie Delta. A large breeding population of Snow Geese, *Chen caerulescens*, is also known on Banks Island. Common Eiders, *Somateria mollissima*, King Eiders, *Somateria spectabilis*, and Old Squaw, *Clangula hyemalis* are abundant along the mainland and Arctic island coasts. Scaup, *Aythya marila*, are also very abundant along the mainland coast and use the rich tidal flats of the major rivers as nesting grounds.

High numbers of ducks and geese are taken annually by Inuit hunters in several Arctic localities. Holman hunters kill a large number of King Eiders at a traditional hunting site near their village in June and July (Smith 1973a). Other traditional duck hunting camps are known at Point Barrow, Warren Point, and Herschel and Baillie islands (Barry 1968, 1973). The Bankslanders harvest a considerable number of Snow Geese each year when the flocks return to Banks Island to nest in early June (Usher 1970).

Mammals

The main cash income of the Inuit hunters is derived from three mammal species: the Ringed Seal, *Phoca hispida*, the Arctic Fox, *Alopex lagopus*, and the Polar Bear, *Ursus maritimus*. Table 1a shows the average price of pelts sold by 16 Sachs Harbour hunters for the years 1963-64 to 1966-67 based on figures from Usher (1970). Table 1b shows the very large increase in the range of prices being paid for pelts in 1974. The current price trends show that mammal furs and skins are becoming an increasingly important source of income for the present day Inuit.

Table 1.

Price of Pelts and Skins.

	(a) Average price of pelts from Usher (1970)		
Year	Ringed Seal	Arctic Fox	Polar Bear
1963-64	$30.00	$24.00	$150.00
1964-65	17.72	14.54	137.03
1965-66	9.16	23.89	150.00
1966-67	9.25	22.44	151.65
	(b) Range of prices for the current year		
1974-75	$25.00-$45.00	$40.00-65.00	$600.00-$900.00

The Ringed Seal has always been the basis of the Inuit coastal economy. Of the Canadian villages being considered, Holman, on western Victoria Island, has had a peak annual catch of approximately 8000 seals which represents one of the highest takes per hunter in the Canadian Arctic. Sachs Harbour on western Banks Island could probably harvest as many seals (cf. Usher 1970, p. 103) but relies to a greater extent on fox for its cash income.

Ringed Seals are distributed throughout the ice-covered areas of the Beaufort Sea. Detailed surveys of the Amundsen inshore and offshore ice indicate that there are a few densely occupied breeding areas and that these are located in large bays or sounds. On the whole the Amundsen Gulf and eastern Beaufort Sea ice is much less densely occupied than the ice along eastern Baffin Island (Smith 1973b). Nonetheless this large area of ice with its relatively low production of seals (Burns and Harbo 1972, Smith 1974) is of prime importance in supplying the seal hunting communities since it is obvious that the few areas of prime breeding habitat could not sustain the present harvest of seals. It is important therefore to delimit the areas of prime breeding ice, especially if they are close to a hunting community since they may provide a significant proportion of the local annual harvest. Recent studies involving the capture and release of marked seals have shown that seals captured at Herschel Island have been shot at Holman and that one seal released at Cape Parry was killed at Point Barrow, Alaska. This supports the idea that seals produced far away from hunting centres are important in supporting the sustained annual harvests. Large-scale movements of seals along the mainland coasts in a westerly direction in the early fall have been documented for several years at at least three different western Arctic netting sites. This redistribution of animals involves mainly yearling and adolescent seals and there is likely a similar spring movement (Usher 1970), but this is more difficult to document.

The other marine mammal species present in the Beaufort Sea are also found in somewhat lower numbers than other Arctic localities, reflecting the generally low productivity in the area. Several hundred Bowhead Whales, *Balaena mysticetus,* move into the area in the early summer from the Bering Sea where they overwinter. In the open-water months these whales are sighted in the Amundsen Gulf and throughout the southern Beaufort Sea. Beginning in September there is movement west along the Alaskan coast and no Bowheads are found in the ice-covered Canadian waters during the winter. The White Whale, *Delphinapterus leucas,* a much smaller species, has a similar seasonal distribution. The White Whales occupy several sites as calving grounds in bays and estuaries at the mouth of the Mackenzie River (Sergeant and Hoek 1974). There are no known summer concentrations of this species along the north slope of Alaska but some may occur around Banks Island. The hunters from Tuktoyaktuk have killed approximately 200 White Whales annually for many years. The maximum number of White Whales in the southern Beaufort Sea has been estimated as approximately 4000 (Sergeant and Hoek 1974).

The Bearded Seal, *Erignathus barbatus,* shows a patchy distribution and is important to the Inuit mainly as a source of leather and protein. Along the mainland coast it is found in relative abundance in such areas as Liverpool Bay and south of Herschel Island. Along the west side of Banks Island, the area between Sachs Harbour and Norway Island contains a relatively high number of this species. The patchy distribution is probably directly related to their dependence on areas of good benthic invertebrate production, where they feed. The comparatively low number of Bearded Seals in the western Canadian Arctic probably also reflects the low benthic invertebrate diversity and production in the Beaufort Sea (Wacasey 1974).

The Walrus, probably the Atlantic subspecies *Odobenus rosmarus rosmarus,* is rarely seen in this area. In the last few years an increasing number of sightings and some catches have been made by hunters from Herschel Island, Holman, and Sachs Harbour. There is some evidence that the walrus was once more plentiful in the area, but was decimated by Bowhead Whale hunters near the end of that commercial fishery in the late nineteenth century.

The Arctic Fox, *Alopex lagopus,* provides from 85% to 90% of the cash income at Sachs Harbour (Usher 1970). Table 2 shows the take at Sachs Harbour for several years and in the other Canadian Inuit villages for the year 1971-72. In both Holman and Tuktoyaktuk the catch of Arctic Fox is also an important part of the winter cash income.

The sizes of Arctic Fox populations are known to fluctuate greatly from one year to the next. The cohort size of the continental population in the Keewatin district, studied by Macpherson (1969), appeared to be regulated by litter reduction and den abandonment in response to the availability of lemming as the main food source. The variation in the size of the breeding population was insufficient to account for the large variation in annual production. In contrast a fox population in West Greenland, which depends mainly on marine mammals and birds since there is a total absence of small mammals, shows no discernable fluctuation in size (Braestrup 1941). It is not clear what the proximate factors are which govern the population numbers of Arctic Fox bordering the Beaufort Sea. The Banks Island population which is the most heavily trapped depends both on lemmings and marine mammals, mainly Ringed Seal, as a food source. The fluctuations in size of catch, which are well documented for this population (Usher 1970), probably are directly related to lemming density and survival of the annual cohort. The continued high reproductive success is also probably due in large part to the availability of a constant source of food in the form of Ringed Seal pups during the three months prior to the breeding season. Tchirkova (1951) indicated that the nutritional state could govern the proportion of breeding foxes and also influence their litter size.

Table 2.

ARCTIC FOX CATCHES FROM SEVERAL INUIT VILLAGES.

	Sachs Harbour	Holman	Paulatuk	Tuktoyaktuk
1963-64	1982			
1964-65	1498			
1965-66	2932			
1966-67	8447			
1970-71	2111			
1971-72	2687	2236	227	2067
1972-73	2000			

My own studies of the Ringed Seal breeding habitat show that Arctic Foxes are important and efficient predators of seal pups in the birth lairs during the months of March to June. The Banks Island trappers believe that there is a seasonal movement of fox off the island to the sea ice in the fall. The dependence of Arctic Fox on the sea ice habitat for food is especially noticeable in low lemming years on Banks Island. During these periods a lower overall catch is associated with a high proportion of the catch being taken in the spring (Macpherson 1960, Usher 1970) when the foxes are returning from the sea ice to the land to breed.

A permissible annual catch of 49 Polar Bears, *Ursus maritimus,* is taken from the four Inuit villages bordering the Beaufort Sea and Amundsen Gulf. No village has failed to take the limit since it was established in 1967. In 1973 there was an increase of 230% in the prices paid for pelts at the western Canadian fur auction over the prices paid in 1971-72 (Jonkel 1974).

The southwest coast of Banks Island from Cape Kellett around to de Salis Bay is a known denning area for Polar Bears. This coastline has been identified as one of 15 major denning areas in the entire Canadian Arctic and is the main area in the western Canadian Arctic (Harington 1968). Most bear kills made by the Sachs Harbour Inuit are in this area. Holman hunters also kill bears in this area as well as further north in Prince of Wales Strait, and further south in the Amundsen Gulf. Other important denning areas also exist on Prince Patrick Island, Melville Island, and along the north coast of Victoria Island. No good data exist yet on the extent of their contribution to the annual catch in the areas to the south.

Preliminary data from three years of Polar Bear tagging efforts by the Canadian Wildlife Service indicate that the Canadian Beaufort Sea population is discrete (Stirling 1974). Bears tagged near Banks Island have been recovered one year later in the same general area.

Although Polar Bears are dependant on the land for denning and cubbing, the greater part of their time is spent in the Arctic marine habitat. Ringed Seals and Bearded Seals in the Beaufort Sea area provide the greater part of their diet. Preliminary calculations on the energy requirement of Polar Bears indicate that one bear would need approximately 120-130 adult Ringed Seals to sustain itself for one year (Oritsland 1974). The availability of abundant food in the form of Ringed Seals, killed in their sea ice breeding habitat, is reflected by a period of rapid weight increase in the spring (Russell 1971). With the reduction of seal availability during the summer open-water period the spring period is probably of prime importance in determining the condition in which the Polar Bear enters the winter. Our studies show that Polar Bears are successful predators of Ringed Seals in their birth lairs. Often the lair is detected, the pup is killed, and the female seal is then caught and eaten. From an energetic standpoint, the killing of the very small Ringed Seal pup alone would not make sense. The adult female seal which is in prime condition provides a substantial meal for the bear. Polar Bears are also successful in killing seals along open leads and in stalking seals which have hauled-out on the ice during the May-June moulting period (Stirling 1975).

Concluding Remarks

The southern Canadian Beaufort Sea is characterized by a general low primary and secondary productivity. Little work has been done in the area and virtually nothing is known about the large regions in the North covered by permanent polar pack. The most abundant species in the higher trophic levels are waterfowl and marine mammals of which the Ringed Seal is the most widespread and abundant. In addition to providing a large part of the Inuit cash income by the trade of seal pelts, it is an important part of the food of Arctic Foxes and Polar Bears, each of which in turn contributes largely to the Inuit cash economy.

References

Alexander, V., *Primary productivity regimes of the nearshore Beaufort Sea with reference to the potential role of ice biota*, in *Symposium on the Beaufort Sea coastal and shelf research*, Arctic Institute of North America, in press.

Barry, T.W., *Significance of the Beaufort Sea coast for migratory birds*, in *Coastal Zone*, vol. 1, Selected background papers (Ottawa: Atlantic Unit, Water Management Service, D.O.E., 1973).

Braestrup, F.W., *A study on the arctic fox in Greenland, Medd. om Gron.*, vol. 131, 1941.

Burns, J., and Harbo, S., *An aerial census of ringed seals Phoca (Pusa) hispida Schreber, along the northern coast of Alaska, Arctic*, vol. 25, 1972.

Grainger, E.H., *Nutrients in the southern Beaufort Sea*, in *Symposium on Beaufort Sea coastal and shelf research*, Arctic Institute of North America, in press.

Harington, C.R., *Denning habits of the polar bear (Ursus maritimus Phipps)* (Ottawa: Canadian Wildlife Service, Report Series No. 5, 1968).

Jonkel, C., *Resumé of the trade of polar bear hides in Canada*, unpublished report, 1974.

MacPherson, A.H., *Mammal abundance — Banks and Victoria Islands, N.W.T. during the summer of 1959* (Ottawa: Canadian Wildlife Service, unpublished).

MacPherson, A.H., *The dynamics of Canadian arctic fox populations* (Ottawa: Canadian Wildlife Service, Report Series No. 8, 1969).

Oritsland, N.A., *Polar bears: ecophysiological model status*, unpublished report, January 1974.

Russell, R.H., *Summer and autumn food habits of island and mainland populations of polar bears — a comparative study*, M.Sc. thesis, University of Alberta, Edmonton, 1971.

Sergeant, D.E., and Hoek, W., *Seasonal distribution of Bowhead and White whales in the eastern Beaufort Sea*, in *Symposium on Beaufort Sea coastal and shelf research*, Arctic Institute of North America, in press.

Smith, R.H., Dufresne, F., and Hansen, H.A., *Northern watersheds and deltas*, in *Waterfowl Tomorrow*, edited by J.P. Linduska (Washington, D.C.: US01, 1964).

Smith, T.G., *The birds of the Holman region, western Victoria Island, Canadian Field-Naturalist*, vol. 87, 1973.

Sergeant, D.E., *Censuring and estimating the size of ringed seal populations*, Tech. Rept. No. 427, 1974.

Stirling, I., *Polar bear research in the Beaufort Sea*, in *Symposium on Beaufort Sea coastal and shelf research*, Arctic Institute of North America, in press.

Tchirkova, A.F., *A preliminary method of forecasting changes in numbers of arctic foxes*, in *Transl. Russian Game Rep.*, vol. 3 (Ottawa: Queens' Printer, 1958).

Usher, P.J., *The Bankslanders: Economy and Ecology* (Ottawa: Northern Science Research Group, Department Indian Affairs and Northern Development, 1970).

Wacasey, J., *Shelf infauna of the Beaufort Sea*, in *Symposium on Beaufort Sea coastal and shelf research*, Arctic Institute of North America, in press.

Wilson, H.P., *Winds and currents in the Beaufort Sea*, in *Symposium on Beaufort Sea coastal and shelf research*, Arctic Institute of North America, in press.

Appendix 5

Environment Canada **Confidential**

Draft 29 August 1973

EPS/DINA Interface Concerns in the Yukon and Northwest Territories*

Background

In the Speech from the Throne, 9 October 1970, when discussing the establishment of the Department of the Environment, the Prime Minister stated:

There is an inherent conflict of interests, however, between those who are seeking the exploitation of non-renewable resources and those who are charged with the responsibility of protecting the environment. This conflict is not irreconcilable, nor is there anything evil in it. Nevertheless, the Government is of the opinion that until such time as environmental values are firmly entrenched these differences are better debated and resolved by ministers in council and not by officials within a single department. For that reason certain responsibilities have been transferred from the Department of Energy, Mines and Resources [to DOE].

He also stated:

There are many departments in the Government which have and will continue to have important responsibilities for the preservation of the quality of our environment. These departments will cooperate with the proposed department of the environment which will have the principal tools to lead the fight against pollution and to help coordinate the efforts of others.

Problem

The above two statements by the Prime Minister establish principles related to responsibilities for environmental concerns and resource development concerns. Recently, the Department of Indian and Northern Affairs has been taking an increasingly active role in the environmental area. While the mandate of the Department of Indian and Northern Affairs is quite clear concerning its responsibilities for promoting resource development there presently exists uncertainty over legislative responsibilities for environmental concerns North of 60. A dual role seems to be built into the Department of Indian and Northern Affairs' mandate because of its responsibilities for promoting non-renewable resource development and for administering environmental legislation e.g., Northern Inland Waters Act, the Arctic Waters Pollution Prevention Act, the Oil and Gas Production and Conservation Act and the Territorial Lands Act. It is evident that the Department of Indian and Northern Affairs now finds itself in the position of making trade-offs which the Prime Minister has suggested "are better debated and resolved by ministers in council and not by officials within a single department." The situation as it presently exists is therefore analogous to the Department of Energy, Mines and Resources' situation prior to the splitting off of environmental responsibilities

*This is one of several internal working papers discussed in Environment Canada during the period when the department was attempting to develop a role in the Northwest Territories and the Yukon. This document's content has not been altered by CARC, except to delete some irrelevant copy (ed.).

from EMR to DOE. Notwithstanding DINA's present environmental mandate the Department of the Environment has been assigned national environmental responsibilities under the Government Organization Act (1971), the Fisheries Act, the Canada Water Act, the Clean Air Act, the Migratory Birds Convention Act, and now the Wildlife Act. Consequently, DOE officials functioning in the Yukon and Northwest Territories are experiencing interface problems relating to fulfillment of their environmental protection role North of 60.

Objective

The objective is to develop a policy of cooperation in environmental matters in the North which recognizes both the responsibilities of DINA and DOE.

Factors

Sections 4 and 5 of the Department of Indian Affairs and Northern Development Act state:

The duties, powers and functions of the Minister of Indian Affairs and Northern Development extend to include all matters over which the Parliament of Canada has jurisdiction, not by law assigned to any other department, branch or agency of the Government of Canada, relating to
(a) Indian Affairs
(b) the Northwest Territories and the Yukon Territory and their resources and affairs; . . .

The Minister of Indian Affairs and Northern Development shall be responsible for:
(a) co-ordinating the activities in the Northwest Territories and the Yukon Territory of the several departments, branches and agencies of the Government of Canada;
(b) undertaking, promoting and recommending policies, and programs for the further economic and political development of the Northwest Territories and the Yukon Territory; and
(c) fostering, through scientific investigation and technology, knowledge of the Canadian north and of the means of dealing with conditions related to its further development.

Sections 5 and 6 of the Department of the Environment Act state:

The duties, powers and functions of the Minister of the Environment extend to and include all matters over which Parliament of Canada has jurisdiction, not by law assigned to any other department, branch or agency of the Government of Canada, relating to
(a) sea coast and inland fisheries;
(b) renewable resources, including
(i) the forest resources of Canada,
(ii) migratory birds, and
(iii) other non-domestic flora and fauna;
(c) water;
(d) meteorology;
(e) the protection and enhancement of the quality of the natural environment, including water, air and soil quality;
(f) technical surveys within the meaning of the Resources and Technical Surveys Act relating to any matter described in paragraphs (a) to (e); and
(g) notwithstanding paragraph (f) of section 5 of the Department of National Health and Welfare Act, the enforcement of any rules or regulations made by

the International Joint Commission, promulgated pursuant to the treaty between the United States of America and His Majesty, King Edward VII, relating to boundary waters and questions arising between the United States of America and Canada, so far as the same relate to pollution control.

The Minister of the Environment, in exercising his powers and carrying out his duties and functions under section 5, shall

(a) *initiate, recommend and undertake programs, and co-ordinate programs, of the Government of Canada, that are designed to promote the establishment or adoption of objectives or standards relating to environmental quality, or to pollution control; and*

(b) *promote and encourage the institution of practices and conduct leading to the better protection and enhancement of environmental quality, and cooperate with provincial governments or agencies thereof, or any bodies, organizations or persons, in any programs having similar objects.*

Presently, therefore, the Department of Indian and Northern Affairs is responsible for both environmental protection and resource development in the North. It is generally accepted practice, however, that government regulatory agencies must operate at arm's length from those requesting licences, e.g. the Canadian Radio-Television Commission, the Canadian Transport Commission, and the National Energy Board. In the case of Panarctic Oils Ltd., DINA provides two members to the Board of Directors: the Deputy Minister, Mr. B. Robinson, and an Assistant Deputy Minister, Mr. A. D. Hunt. There is some question if DINA can effectively carry out its environmental regulatory responsibilities contained in the Northern Inland Waters Act, the Arctic Waters Pollution Prevention Act, and the Territorial Lands Act and also be a member of the Board of Directors of a large resource development concern such as Panarctic. The principle of regulatory agencies operating at arm's length seems to be in jeopardy.

DOE has a duty in the Territories to ensure that adequate environmental protection in an overall sense is carried out. Also several specific environmental protection functions are the sole responsibility of DOE, e.g. the functions under the Clean Air Act. Concessions to resource development at the expense of the environment should not be made within forums, such as the Territorial Land Use Committees and the Territorial Water Boards, without adequate provision for consideration of environmental effects. It is the responsibility of DOE to advise DINA about environmental effects. In certain instances problems have developed which could have been prevented had DOE inputs been adequately provided for.

This situation was evident in the application for a water use licence by the Northern Canada Power Commission (a Crown Corporation) for the AISHIAK Power Project (Yukon). This crown corporation reports through the Minister of Indian and Northern Affairs. The licencing agency for the water use licence was the Yukon Water Board chaired by the Regional Manager of Lands, Waters, and Forests of the Department of Indian and Northern Affairs which reports to the Minister's office. The permit approval was signed by the Minister. The NCPC Environmental Impact Study was undertaken by consultants in two phases and included a four- to five-day duration botanical survey under fresh snow cover!

Another area of concern is that present DOE involvement in the Advisory Committee on Northern Development (ACND Committee) chaired by the Deputy Minister of DINA and the sub-committees may not be at a senior enough level and continuity in representation has not been provided for in all cases, e.g. DOE participation in the ACND Coordinating Committee.

In a memo dated 25 April 1973 the Chairman of the DOE Northern Regional Board stated that "Canada's northern policies seek to satisfy social and material expectations inside and outside the North, and clearly depend for implementation on a process of balanced decision making based on appropriate data and advice." He went on to cite the major areas of immediate concern.

Clarification is required with DINA as to its role as the prime contact with industry, a mandate given that Department by Cabinet. DOE cannot possibly, and does not in practice in the provinces, discharge its responsibilities without direct consultation with industry.

Another problem is of even greater concern to Regional Directors. In essence, it is that our Department is publicly seen to be an advisor to the resource development arm of the Department of Indian Affairs and Northern Development, yet in reality our advice can be ignored or rejected. This situation is one which may lower public trust in the Department and the commitment of the Government to good environmental management in the North. Some of the specifics of the problem are: the Department of Environment has been represented on the Land Use committee in Yellowknife since November 1971, and does have the opportunity to comment on the permit applications presented to that body prior usually to permits being issued, but if (as is frequent) a particular company wishes to change the stipulations of the permit after an operation begins, the permit is usually not brought back to the Committee but may be changed in the field by the DINA Engineer. For example, one permit stipulated supplies were to be moved from barge to drill site by helicopter. By field amendment the operation became one of dumping supplies on shore and hauling them with tracked vehicles over thawed ground (in August). DOE advises only on request, and DINA may request or not according to its judgement.

Other examples are:

(a) On 26 July 1973 an emergency meeting was called by the Chairman of the Arctic Waters Oil and Gas Advisory Committee of the Northwest Territories. This meeting was called to discuss Land Use Application no. N73-J419 and Land Use Application no. N73-A516. Both applications were submitted by Imperial Oil Co., the first to construct a second off-shore island in the Beaufort Sea, and the second to drill from that island. The short notice given to the committee members made it possible for only two to attend, in addition to a substitute for the Chairman of the Committee. These two were the EPS representatives and the DINA Oil and Gas Division representative, both from Yellowknife.

It appeared that Imperial Oil wished to commence construction of the island immediately, and to drill from it on or about 1 December 1973. The EPS representative was not prepared to sanction either permit and in fact, was reticent to discuss them, because the terms of reference for the AWOGA Committee had not yet been provided by DINA to the committee members, nor had Cabinet yet given approval in principle for off-shore activities. Also, no one else from DOE had yet seen these applications or commented on them. The EPS representative stated however, that DOE personnel would be prepared to review and comment to the NWT Land Use Committee on the island construction application by 15 August 1973.

On 9 August 1973, the Chairman of the NWT Land Use Committee issued permits to Imperial Oil not only to construct the island, but to drill from it as well. This action has been completed with no input from DOE despite the fact that the Land Use Committee Chairman, who is also the Chairman of the AWOGA Committee, was well aware that input from DOE was to be presented by 15 August 1973, some six days later. This example illustrates the lack of acceptance of DOE's role in the North and the lack of more than superficial concern about the environmental aspects of development.

(b) Air transportation facilities in the Arctic are being developed by MOT for DINA (50 landing strips proposed), and each airstrip is then approved by the Regional Manager, Water, Forests, Land and Environment Division of DINA who administers the Territorial Land Use Regulations.

(c) Similarly, the Mackenzie and Dempster Highways are being constructed by DPW under contract to DINA in both NWT and the Yukon. This is another example of DINA proposing, funding, and approving from an environmental viewpoint, a northern development project. DINA has demanded that DOE not deal directly with DPW on environmental concerns relative to the Mackenzie Highway. DINA cannot demand this if DOE is to carry out its role.

Land Use Advisory Committees (LUAC) operate in the Yukon and Northwest Territories under the Territorial Lands Act and its Regulations established by DINA, and provide an administrative mechanism for receiving advice relating to the issuance of land use permits. The Territorial Land Use Regulations are presently the most potentially powerful and all encompassing set of regulations for environmental control in existence today in the North, yet there is no requirement that DOE views be accepted on permit applications. These committees have been in operation for over two years and have as yet to develop terms of reference for their operations despite several requests by DOE. Decisions on land use permits can have serious environmental consequences and there is evidence of developmental interests receiving consideration without sufficient consideration of environmental factors.

Another example is a recent application made to the Land Use Committee for the Northwest Territories by Eureka Exploration Ltd. to carry out seismic work on the east coast of Baffin Island. This area is well beyond the current land use zone administered by DINA under the Territorial Lands Act. There is a question as to whether the explosive permit authority of the Fisheries and Marine Service of DOE would not be more appropriate to these off-shore seismic activities.

There presently exists a possibility for duplication in the environmental protection provisions of Section 33 of the Fisheries Act and the Northern Inland Waters Act. Because of the DOE environmental protection responsibilities contained in the Fisheries Act, periodic sampling of effluents, in particular bio-assays, must be conducted by this Department. Monitoring and surveillance programs under the Fisheries Act are required and a capability presently exists within the Environmental Protection Service (EPS) to carry out these responsibilities within DOE.

Under the Northern Inland Waters Act provision is made for a Water Licence Board to which the Department of the Environment provides one member in an advising capacity only. Although DOE has responsibility for the Fisheries Act it is only one voice within the Water Board forum and can be heeded or disregarded. At present water use licences have no explicit provision that the Fisheries Act must be adhered to which creates difficulty in subsequent enforcement situations. The time frame presently employed by the Water Boards does not allow sufficient time for adequate assessment of environmental concerns by the Department of the Environment. A recommendation by the Department of the Environment representative to the Water Licence Board of the Yukon Territories to restrict initial issuance of licences to five years looks as if it might be adopted. In the case of the Yukon Water Board, activities have been divided among sub-committees (e.g. mining, power development, municipalities, etc.) further weakening the influence

DOE can bring to bear within this forum. The DOE Water Board member is on only one of these sub-committees.

The Arctic Land Use Research Program (ALUR) administered by DINA finances studies on pollution control technology at universities, e.g., containment of mining wastes, bacterial degradation of oil, use of tracked vehicles on Arctic terrain. Technology development has been identified as an EPS function, yet there is no mechanism for DOE to set terms of reference for or monitor the work. It is interesting that ALUR currently funds approximately $80,000 of applied research work by the Arctic Technology Section of the EPS Northwest Region.

ALUR acts as an environmental advisor to Water, Forests, Land and Environmental Division of DINA for approvals of land use permits and water licences, for the Mackenzie Highway, and for the proposed Mackenzie Valley Pipeline. DOE also has been specifically designated as an advisor on these projects.

A recent document prepared by DINA relating to off-shore drilling further illustrates their dual role of encouraging development activity and at the same time attempting to assume responsibility for enforcement of permit conditions. Recent branch and lower organizational name changes which have taken place within DINA re-enforce the belief that the Department is actively expanding its activities into environmental protection areas of the North. These changes are contributing to confusion in the Territories because of the use of similar terminology by two federal government departments. Examples of name changes are as follows: from Northern Economic Development Branch to Environment and Natural Resources; from Water, Forests and Land Division to Water, Forests, Land and Environmental Division; from Resource Management Officers to Environmental Protection Officers.

An advertisement in the *Edmonton Journal* dated 21 June 1973 explained the duties of an Environmental Control Engineer (Mining): as "to plan, organize and co-ordinate environmental surveys in mines including ventilation, dust count, radiometric, roaster track emission, noise level, mining atmosphere, effluent and water quality."

Alternatives

There are three alternatives:

(a) Legislative amendment assigning all environmental responsibilities to DOE. DOE could press for legislative amendment to separate environmental and developmental responsibilities between DOE and DINA in an analogous fashion to the DOE/EMR division of responsibilities. This solution may be required in the long term but it would require a good deal of time and full agreement by Ministers. This course of action should only be pursued if all else fails.

(b) Full reliance by DOE on DINA to carry out environmental protection mandate in the North.
This alternative is certainly preferrable to any arrangement which fosters bureaucratic competition about environmental matters in the North. It has the disadvantages of not utilizing the expertise available in DOE and would result in environmental trade-offs being determined within a single department. DOE also has legislative requirements for which the Minister is fully responsible and which cannot be delegated to another Department except by legislative amendment.

(c) Modification of licencing mechanisms and harmonization by DOE and DINA of their activities in the North. This alternative could be worked out at the Deputy Minister level or if necessary at the Ministerial level on a bilateral basis. The time frame for making changes would be shorter than for alternative (a) and (b) above. The solution would take advantage of expertise, experience, and resources of both DINA and DOE and provide for each Department to carry out its responsibilities in a meaningful way. A full understanding and acceptance of each other's roles and objectives would be necessary and continued co-ordination in the future is essential at all levels. Because of these considerations this is the preferred alternative.

Financial Considerations

The role of DOE must be further defined before financial implications may be clarified.

Federal-Territorial Relations Considerations

It is essential that the federal government act in a consistent and harmonious manner in the Territories.

Interdepartmental Considerations

There are serious interdepartmental considerations involved which directly relate to the responsibilities of DINA and touch on the responsibilities of Energy, Mines and Resources and the Ministry of Transport. Additionally, other services within DOE may also have substantive problems and before this matter can be meaningfully discussed with other government departments, other problem areas within DOE should be identified.

Public Relations Considerations

It is essential that duplication be minimized North of 60, and that federal government departments operate in a harmonious, supportive fashion.

Conclusion

In summary then, confusion appears evident between responsibilities of DINA and DOE for the functions of resource development and environmental protection. The principle of regulatory bodies operating at arm's length from the interests being regulated is not being respected in the case of Panarctic Oils Ltd. Harmony between DINA and DOE responsibilities and priorities in the North must be further developed. The ACND committee structure and the ICE committee structure may provide a forum for accomplishing this. For example the re-alignment of the Environmental Social Committee of ACND under the Interdepartmental Committee on the Environment may now be timely and sensible. There is a potential for duplication in the monitoring and surveillance functions associated with requirements of the Northern Inland Waters Act, and the Fisheries Act. The capability for prosecuting delinquent persons or agencies must be provided for by DOE and DINA in a co-ordinated and effective way. The Arctic Waters Pollution Prevention Act, as it is presently being administered, does not provide for any DOE involvement. The recent DINA cabinet document on off-shore drilling does not provide for full DOE approval of off-shore drilling permits. There is a lack of provision for effective input by DOE into the decision making process related to land use licences. There is also potential for confusion in recent organizational re-designations within DINA and organizational designations within DOE.

The present situation provides a potential for bureaucratic competition, duplication of activities, unclear responsibilities, uneven application of environmental protection legislation, and a fundamental conflict of interest North of 60 between resource development and provision for protection of the environment.

The interface problems which presently exist are serious and will be difficult if not impossible to resolve at board and regional levels. While the examples cited are based on EPS experience other DOE services may also be experiencing serious interface problems. It is essential to consider the subject of DOE-DINA interfaces in a departmental context before any discussions of the problems with DINA take place.

Recommendations

It is recommended that:

(a) discussions within DOE and subsequently with DINA take place at a senior level concerning environmental responsibilities North of 60 with a view to harmonizing responsibilities and activities.

(b) a joint DINA/DOE review should be undertaken of administrative arrangements and approval criteria for permit issuance and amendment with provision for subsequent enforcement as required in view of legislative responsibilities in the North.

(c) there be established provision for environmental review and amendment for major undertakings prior to land use or water use decisions.

(d) DOE be made a full partner of the Territorial Land Licencing System and official approval of applications by both DOE and DINA be provided for.

(e) time-frames for review of permit applications be sufficient to allow adequate environmental consideration of applications.

(f) DINA be requested at senior level to propose terms of reference for Land Use Committees and Arctic Waters, Oil and Gas Committee.

(g) utilization of both the explosive and the land use systems under the authority of the Fisheries Act and the Territorial Lands Act for off-shore seismic activities North of 60 be studied.

(h) the Territorial Water Boards provide for joint approval by DINA and DOE of applications.

(i) DINA and DOE monitoring and surveillance functions required under the Fisheries Act and the Northern Inland Waters Act be harmonized with provision for full exchange of information and audit.

(j) for off-shore drilling permits, DOE and DINA have a joint approval responsibility with provision for DOE environmental assessment for particular environmentally sensitive projects.

(k) DOE review its involvement and representation to ACND Committees to provide for appropriate levels of representation, continuity, and effective internal communication of activities of these committees. The relationship of the ACND to ICE merits definition.

(l) communication lines within DOE regions and headquarters be strengthened, recognizing that the regional responsibilities are associated with the project by project interface with DINA and the headquarters responsibilities are associated with the policy and program development aspects.

(m) consideration be given to dealing with the matter of DINA participation in the Board of Directors of Panarctic.

Appendix 6

The Probable Nature and Behaviour of Oil Spills in the Beaufort Sea, and the Feasibility of Cleanup

G. D. Greene and D. Mackay*

Exploration for oil and gas in the sedimentary basins of the Canadian Arctic is concentrated in two regions, the Mackenzie Delta-Beaufort Sea area and the Arctic Islands. Modest oil finds have already been made onshore in the Mackenzie Delta but the most promising areas for exploration are offshore in the Beaufort Sea. In 1973, the first step toward offshore drilling was taken when exploration for oil began from artificial islands built in approximately 10 feet of water in the Beaufort Sea. Plans are now being made to begin drilling in waters north of the Mackenzie Delta from offshore rigs. Aspects of the Canadian federal government response to this situation have been described by Logan (1975) and Ross (1975).

This review is an attempt to assemble and comment on the existing information on Arctic oil spills on water and ice. The emphasis is on the probability, nature, and consequences of a major oil spill in the Beaufort Sea, resulting from an oil well blowout during exploration.

It has been reported that Dome Petroleum of Calgary will begin exploratory drilling in 1976 some 100 to 200 km offshore in the Beaufort Sea, north of Kugmallit Bay in 15 to 60 m of water (*Oilweek*, 8 July 1974 and 7 October 1974). Each well is estimated to cost about $20 million. This venture will be the first attempt at offshore drilling in Arctic ice-infested waters, although Panarctic in the winter of 1974 and again in 1975 drilled successfully from a stationary reinforced ice sheet 18 km offshore from Melville Island in the Arctic Archipelago.

Drilling will take place from one or two ice-strengthened 358-foot drillships. The ships will be connected to eight buoys anchored to the sea bed. Using this system, wells up to 25,000 feet deep can be drilled in water depths up to 1000 feet. The fixed mooring system will be designed so that the ships can be released and re-connected to the well rapidly in the event of having to move because of ice intrusion. The release time is expected to be about six hours (*Oilweek*, 7 October 1974). Such a situation is quite likely to arise due to the highly dynamic nature of the Beaufort Sea ice pack as discussed later.

Dome plans to drill during an open water period of 110 days from July to October (*Oilweek*, 8 July 1974). Past experience suggests that this estimate may be optimistic. For example, in 1974, Dome had to cancel its Beaufort Sea program due to adverse ice conditions caused by a strong prevailing north-westerly wind during most of the summer. Dome had planned to begin well locations in the summer of 1974 but the pack ice remained within 50 km offshore north of the Mackenzie Delta and did not permit any ships to enter the proposed drilling area.

*Department of Chemical Engineering and Applied Chemistry, and the Institute for Environmental Studies, University of Toronto. Reprinted by permission.
The authors wish to express their gratitude to the Committee for Arctic and Subarctic Research of the University of Toronto for financial support; to the National Research Council for a scholarship to G. D. Greene; and to the many individuals who provided valuable opinion and information during the compilation of this review.

The possibility of a blowout occurring is always present during drilling, particularly when wild-cat wells are being drilled offshore. The causes and characteristics of such blowouts are discussed later. The unique offshore drilling conditions in the Beaufort Sea contribute several factors which make drilling relatively more hazardous than in temperate waters. Most important is the possibility that ice floes may drift in on the drill ship while drilling, either forcing the operators to disconnect if there is time, or if not, breaking the drill stem by collision with the vessel. There is a possibility that drilling may be attempted by trying to keep open water clear around the drill ships if encroachment of ice occurs. Drilling in such conditions is difficult because of the rapidity with which ice floes can advance and the large momentum which they may carry. Dome plans to monitor ice floe movement by radio and if necessary tow away or attempt to break up floes using ice-breaking work ships. Despite these precautions, the probability of a drill stem break is judged to be relatively higher in the Beaufort Sea than in temperate waters.

Recognizing these hazardous conditions, Dome has developed a two-stage blowout preventer (BOP) which will allow quick disconnection of the marine riser (that portion of the drill stem in the water) either at the bottom of the ocean or above the bottom. If the drill ship must break connection, the upper BOP stack will close off the well leaving the marine riser disconnected. Hopefully, this will ensure that there will be no blowout. There is also a possibility that a very deep draft ice floe could enter the drill zone and damage the drill stem or BOP system. The entire blowout prevention system is new and relatively untried although the upper BOP stack lying on the ocean floor has been used by Panarctic in its ice sheet well drilling (*Oilweek*, 8 April 1974 and 7 October 1974). It is clearly essential that break-off can be accomplished in a real situation in a short period of time.

Blowout can also occur during inplace drilling, as has occurred in temperate drilling situations, and closing off the well may then be very difficult. Panarctic in its Hecla ice sheet well had a hydraulically activated BOP activation system remote from the rig. In an offshore situation this would require that another ship be in place in open water drilling.

A second factor which could lead to drilling problems is the presence of very high formation pressures in the Mackenzie Delta. Imperial Oil's first artificial island well (Immerk) and Sun Oil's artificial island well (Pelly B35) were both abandoned because of high formation pressures where there was fear of formation fracture (*Oilweek*, 10 February 1975). It is of course standard procedure to monitor wells closely for signs of high pressure zones and Dome plans to use enlarged casings to allow pumping down excess weighted drilling mud to balance the high pressures.

A third factor which increases the hazards of drilling in this area is the weather which is discussed in more detail in a subsequent section. The severe Beaufort Sea storm of 1970 which produced winds of 50 knots for a period of three days and caused considerable coastal damage indicates that short periods of intensely bad weather can be expected in this region (Brown 1971). The performance of drill ships under these conditions is almost unpredictable. Certainly experience in Hudson Bay, in which a well was abandoned by Aquitaine in 1969 and a drill barge was badly damaged by ice and wind, indicates that weather must be taken into account in any offshore drilling activities.

It is suggested that these three factors — ice, high formation pressures, and weather — will increase the probability of an offshore blowout and that it is prudent to examine these factors closely and assess the consequences of such an accident and the degree of preparedness for cleanup.

Ice and Weather Conditions

A detailed understanding of ice and weather conditions in the Beaufort Sea is clearly of critical importance in the proper planning of offshore drilling operations. The ice can be considered as being of three types (Rogers 1974, Dunbar and Wittman 1962, Rosandhaug 1974).

First is shorefast first-year ice which is subject to break-up each year by thawing and tide or current-induced motion. This ice has a relatively high salt content of several thousand ppm, and is interlaced with brine channels formed in the ice as concentrated brine is drained down through the ice during freezing. These make the ice relatively porous and may provide continuous channels from the bottom to the top of the ice as the ice thaws. Second is pack ice which may be one or more years old and is subject to short-term wind-driven motion and long-term current and atmospheric-induced movement. It is characterized by a rough surface and subsurface pressure ridge keels. Multiyear pack ice is harder than first-year ice, being relatively salt free, the brine having drained out during previous summer thaws. As a result, even when it warms up in summer, multiyear ice does not contain continuous channels. Third is the shear zone ice which lies in the region between the first two categories.

Until now, drilling from artificial islands has taken place in the shorefast zone where ice conditions and movement are fairly predictable. However, plans for offshore drilling in 1976 include operations outside the shore ice zone which generally extends 50-70 km from the shore into the shear zone and to the edge of the pack ice which may begin from 80 to several hundred km offshore (Markham 1975, Ramseier 1975).

The pack ice in the Beaufort Sea is subject to a continuous clockwise drift known as the Beaufort Gyre, in which portions of the pack ice, often moving together but sometimes expanding or contracting, move at speeds of one to five km per day. Movement is rarely linear and is often erratic in direction (Dunbar and Wittman 1962).

The shear zone is a dynamic area in which enormous forces are experienced in winter, spring, and fall as the moving pack ice encounters fixed shorefast ice. In the summer short-term movement of the edge of the ice pack caused by winds and currents may cause portions of the ice to move distances of up to 60 km in one day (Markham 1975). Currents are generally weak in the area north of the Mackenzie Delta and wind speed generally controls ice movement at approximately 2% of the wind velocity. Depending on prevailing summer wind conditions the front of the ice pack may be from 30 to 300 km offshore north of Richards Island and Kugmallit Bay at the end of the thaw season.

Weather conditions in the Beaufort Sea during the late summer and early fall are considered to be relatively unstable. Storms due to frontal activity in the Bering Strait and formation of pressure lows in the Mackenzie Valley are not uncommon in the summer (Markham 1975). The 1970 storm which produced winds of 50 knots and lasted three days caused considerable coastal damage. Wind-induced movement of ice into the Mackenzie Bay was at a rate greater than 40 km per day. This storm, although exceptionally severe, was not rated as a 100-year storm (Markham 1975).

It is thus apparent that the area planned for offshore drilling is subject to quickly changing and hazardous weather and ice conditions which must be regarded as significantly different from those encountered in established offshore oil exploration areas.

The Nature of a Blowout Incident

"The drilling of wildcat wells in unknown geological formations can be particularly hazardous" (Kash et al. 1973): so stated a detailed technology assessment of offshore drilling in the United States. Blowouts usually arise from one of two situations. First is penetration of a high-pressure formation during drilling when insufficiently weighted drilling mud can be pumped into the well to balance the underground pressure. Second is the mechanical failure of the drilling mechanism. Such failures usually occur during lifting or lowering of the drill string. Analysis has shown that the major causes of blowouts are human error and storms. A 1971 study by Kennedy (Kash et al. 1973) showed that 65% of the blowout incidents recorded occurred during production under conditions when well performance is, or should be, much better controlled than during exploratory operations. Human error and storms were the causes.

In an analysis of offshore accidents in the U.S., Kash et al. (1973) concluded that in about 14,000 wells drilled, there were 37 blowouts, in 15 of which oil was released, totalling about 290,000 to 1,100,000 barrels. There were also 56 deaths, about 100 injuries, and 20 fires. Another 11 undocumented blowouts are believed to have been caused by Hurricane Hilda in 1964.

This accident rate of 0.2% compares with 0.04% for blowouts on land, the factor of 5 presumably reflecting the more adverse working conditions offshore.

It is suggested here that the probability of a blowout in the Beaufort Sea will be increased by the following factors:

1. The cold weather conditions and isolation which make work more difficult and increase the risk of human error. This effect is common to all Arctic operations.
2. The presence of high formation pressures which may lead to an uncontrollable situation.
3. Ice invasion causing rapid drill stem release and possible breakage.
4. Severe open water storms damaging or sinking a drill ship and possibly resulting in breakage of the drill stem.
5. Ice scour causing a well which has been capped after a successful breakoff to be damaged.

It is estimated that about 375 wells have been drilled in the Mackenzie Delta and Beaufort Sea and another 100 wells in the Arctic Islands (*Oilweek*, 3 March 1975). There have been two blowouts (both Panarctic gas wells) giving a failure rate of about 0.4%. While it is recognized that these occurred in the early stages in the Arctic Islands drilling, several well drilling stoppages have occurred due to high formation pressures and it is thus concluded that a blowout rate of 0.5% to 1% is reasonable for Beaufort Sea offshore drilling.

Information obtained from two previous oil blowouts in the Santa Barbara channel and Platform Charlie in the Gulf of Mexico, and two gas blowouts by Panarctic in the Arctic Archipelago, illustrate the likely magnitude of a Beaufort Sea oil blowout and the difficulties involved with establishing control of a well.

The Platform Charlie blowout in February 1970 resulted in an unchecked flow of oil from the ocean bottom at a rate of 1000-3000 barrels per day with a total amount spilled estimated between 35,000 and 65,000 barrels (Alpine Geophysical 1971). This blowout burned for one month and then flowed oil unchecked for three further weeks.

The Santa Barbara blowout resulted in the release of about 5000 barrels per day of crude oil (University of California, Santa Barbara 1972). Much of the flow was checked after 10 days by drilling a relief well from an adjacent platform which was previously in place. Leakage from a seabed fracture continued at upwards of 200 barrels per day for 90 days. The total amount of oil spilled has been estimated to be about 93,000 barrels although other estimates vary from 78,000 to 780,000 barrels (*Oil on the Sea,* 1971).

Panarctic experienced gas blowouts during wildcat drilling in the Arctic Islands in 1970. The first well flowed water and a small amount of gas for over 12 months during which several unsuccessful attempts were made to drill a relief well. The well was finally capped 16 months after the initial loss of control. The estimated flow of water from the high pressure zone was 50,000 barrels per day (*Oilweek,* 20 March 1972). The second blowout occurred in a high-pressure gas-bearing zone at 2000 feet and burned, fed by an uncontrolled flow of an estimated 42 million cubic feet of gas per day (*Oilweek,* 16 November 1970). This is equivalent to a high production rate gas well. An equivalent oil production rate would be several thousand barrels per day; using an energy equivalence of 6000 cubic feet to the barrel this rate is 7000 barrels per day.

It is noteworthy that severe weather conditions in Hudson Bay forced Aquitaine to abandon an uncapped well in 1969. Fortunately in this case, no oil or gas was encountered (Pimlott 1974).

The only methods presently available for handling a blowout are to ignite the well, or to drill a relief well which reduces the pressure in the damaged well and allows it to be capped. Given the highly variable nature of Beaufort Sea pack ice, it is entirely possible that it could take a year before conditions would allow a ship to enter the area and drill a relief well. Ignition of a blowing oil or gas well could be very difficult or impossible under heavy ice conditions.

Lester and Beynon (1973) have reviewed spillage estimates and have described the North Sea approach to this problem. Brockis (1974) indicated that the North Sea contingency plan allows for up to 15,000 barrels per day for one week and 100,000 barrels in total. A request for tenders for a research contract for part of the Beaufort Sea Study quoted oil flow rates of 2000 barrels per day as a basis for study (DSS 1974).

Based on the above incidents, it seems fair to assume that an oil blowout in the Beaufort Sea (where formation pressures are known to be high) could result in the release of from 2000 to 5000 barrels of crude oil per day. Flow would likely continue unchecked for at least one month by which time it is possible some relief measures could be taken to reduce the flow to perhaps 1000 barrels per day. Flow at this rate could continue for about a year. It is thus possible that the blowout incident could release 60,000 to 150,000 barrels of crude oil in the first month and a total of 100,000 to 600,000 barrels during the entire incident.

The potential consequences of a spill of this magnitude into the Beaufort Sea, or even of a spill an order of magnitude smaller, are unknown at present. It has been suggested that such spills could lead to significant damage to marine life and migrant birds, disruption of the Beaufort Sea ecosystem, melting of large portions of the ice pack due to the spreading of the oil into the Beaufort Gyre, and even lead to climatic changes (Pimlott 1974, Ramseier 1972, Campbell and Martin 1974).

Although there has been some experience in handling oil spills under cold weather and ice-infested conditions, for example, with the *Arrow* tanker incident in Chedabucto Bay or the various minor oil spill incidents in the Gulf of St. Lawrence and the Canadian Arctic, it is clear that an offshore blowout in the Beaufort Sea comparable in volume to the Santa Barbara incident will require the implementation of clean-up measures substantially different from anything used in the previous treatment of oil spill emergencies. To be prepared to treat such incidents effectively requires the development of a new cleanup technology which must be based on a deep understanding of the behaviour of the oil.

The Physical Behaviour of Spilled Oil

Oil Rising and Spreading in Arctic Waters

Since all drilling is planned for substantially ice-free conditions, the first situation which must be considered is a spill in open waters.

A blowout in such circumstances would result in oil rising through the water column to the surface in the vicinity of the well. The oil in the well may be at a temperature of about 70°C. It is believed that the oil will be cooled almost to ambient water temperature during the rising process and will be dispersed into small droplets by the time it reaches the surface. It is also likely that with the oil will be an amount of natural gas which will rise as gas bubbles. During its rise through the water, the oil will heat the water and may cause a substantial rising current due to both density changes and fluid drag. It is also likely that in the dispersion process, a considerable amount of the oil will be broken down to droplets of diameter 1-20 microns, which too small to rise with an appreciable velocity, will be carried by the ocean currents and will remain as dispersed oil particles until they are ultimately destroyed by dissolution, evaporation, and biodegradation. A considerable amount of the more soluble components of the oil will dissolve in the water and it can be expected that dissolved hydrocarbon concentrations of the order of 1 to 30 mg/l will be encountered in the water column, together with a comparable concentration of emulsified hydrocarbon. These hydrocarbons are thus present at levels which can have significant toxic effects on marine biota (Moore and Dwyer 1974). The dispersion of an oil column rising in water has been noted in laboratory studies by Keevil and Ramseier (1975) and by Bell (1974).

As the oil droplets reach the water surface it is likely that they will coalesce into an oil slick which will then spread on the water surface. This spreading process has been studied extensively and it has been shown that the fluid dynamics are controlled initially by a balance between inertial and gravity forces followed by viscous and gravity forces with surface tension forces finally controlling the spreading as the slick becomes very thin. Several correlations have been proposed and tested to estimate the area of the slick as a function of time and spill volume.

Studies by Glaeser and Vance (1971) and by McMinn and Golden (1973) in Alaska have shown that oil spreading on sea water at -1°C behaved similarly to spills on more temperate waters, passing through the same force regimes. Glaeser and Vance observed a net negative spreading coefficient for the oil on water below 0°C although this may be a property of the oil. It appears that oil spills on Arctic water have a tendency to stabilize or self-contain at a relatively greater thickness than occurs in temperate regions where oil spreads to form very thin, even mono-molecular layers on the sea surface. Estimated ranges for oil thicknesses on near freezing waters vary from 0.1 to 1.0 cm (Ayers et al. 1974). The total physical behaviour of an oil slick on cold water is very complex and changes will take place in the oil viscosity, volatility, and density as dissolution, evaporation, and possibly water-in-oil and oil-in-water emulsion formation take place. Wind will also drive the slick, and sea surface roughness, with white-capping present will profoundly affect the slick behaviour especially with regard to emulsion formation. The stable water-in-oil emulsions (containing up to 50% water) observed in Chedabucto Bay had very high viscosities and proved difficult to handle. They may also be susceptible to dispersion throughout the water column.

In recent years, studies of the restoration rates of oil-stressed environments have shown that one of the most severe long-term oil spill effects has been the incorporation of oil into sediments where there is a severe and lasting effect on benthic organisms. The mechanism of oil sinking is far from clear but is thought to be largely by adsorption on falling inorganic particulates in the water column, by oil droplets being transported by currents, as well as increase in oil density beyond that of water due to weathering. If bottom conditions are anaerobic then oil degradation may be very slow. It has been reported that depressions or "black holes" exist on the sea bottom in which anaerobic conditions may exist. Oil has been tentatively reported but not verified in these "holes" (Wickland 1974). Unfortunately, there is a high sediment load in the Mackenzie River and Delta and it seems likely that a quantity of oil will be sedimented in the Beaufort Sea possibly leading to a long-term benthic contamination problem. The biological consequences of this effect are not known. It should be noted that the extent of oil dispersal in the water column will be much greater when oil is released at the ocean bottom than when it is released on the surface.

The slick will also be subject to wind movement and will travel at approximately 3% of the surface wind velocity. Some components of the crude oil will dissolve in the water, particularly the aromatics, although the amounts dissolved will generally be small in comparison to the amounts evaporated. These rates have been considered recently by Leinonen and Mackay (1975) who have suggested that for all but the polynuclear aromatics the evaporation rate will exceed the dissolution rate by a factor of 50. In the case of alkanes this factor is closer to 1000.

In conclusion, it is probable that the physical behaviour of an oil slick in ice-free Arctic waters will not be significantly different from behaviour in temperate waters, provided that allowance is made for the high viscosities and lower volatilities resulting from lower ambient temperatures. However, slick thicknesses may be greater. Only when the oil properties, sea conditions, and weather are known can the oil behaviour be predicted with any accuracy. There is still much ignorance about the behaviour of the rising oil.

Oil on Sea Ice

Some oil released from a blowout may reach the ice top surface by flowing directly from the rig or by welling up through a hole in the ice. The behaviour of oil on ice has been considered by Glaeser and Vance (1971) and on a small scale by Chen et al. (1974). Glaeser and Vance were unable to specify an empirical model for oil flow on ice because of the roughness and high degree of oil absorption by the summer ice studied. They established that spreading on ice was substantially independent of temperature and the physical properties of the crude oil. The same conclusion was reached by McMinn and Golden (1972) who found that ice surface roughness restricted spreading. They determined an average roughness height for sea ice as three cm, excluding, of course, large formations such as pressure ridges. They concluded that slick size was primarily dependent on surface roughness. In that study, however, there was little absorption of oil into the ice surface. The spread of hot oil on ice reduces the extent of absorption by causing ice melting and re-freezing.

In a laboratory study, Chen et al. (1973) derived equations relating the size of an oil slick to the volume spilled and time. This, however, was on smooth ice performed under laboratory cold room conditions and the results cannot necessarily be scaled up to large spill conditions.

The studies have indicated that the dominant mechanism for oil flow on ice is gravity viscous flow and that surface tension is unlikely to play a significant role. The thickness of oil on a smooth ice surface will probably reach a final value of 0.3-2 cm. On rough ice the surface will be saturated with oil and above this will lie a corresponding layer of oil. This model provides an approximate basis for calculating the area affected by oil spilled on ice.

Snow will have significant effect on the behaviour of oil on ice, the snow being either in place prior to the spill or blown onto the site after the spill. McMinn (1972) observed that snow blowing across the surface of a crude oil spill on ice had a tendency to stick to the oil and gravitate downwards into it. The resulting mixture contained up to 80% snow and took the form of a relatively "dry" mulch, as long as the ambient temperature remained below the pour point of the oil. The penetration of the oil into the snow has been considered in some detail by Mackay et al. (1975) and approximate calculations can be made of the extent of oil penetration, provided the oil viscosity and snow porosity are known. Observations in Arctic conditions have shown that a free oil surface will rapidly accumulate blown or fallen snow forming such a mulch and that under normal conditions, a blanket of snow will obscure the oil on top of the ice. This will make detection of the oil difficult. A remote sensing technique based on temperature may be feasible.

Oil under Ice

If a blowout occurs under winter conditions or continues into winter, it is likely that the oil will rise to the undersurface of the ice and will spread out along the ice-water interface. Studies of the spreading of oil under ice have been reported by Glaeser and Vance (1971) in Alaska, by Greene and Mackay (1975) in a fresh water pond in Ontario, by Bell (1974) at Resolute Bay, and studies were undertaken in the winter of 1974-75 as part of the Beaufort Sea study (Brown 1975).

Oil spreading under ice will be largely controlled by gravity-induced buoyancy of the oil and resisted by the viscosity of the oil and the drag of the water. Laboratory observations of oil droplets at an under-ice surface by Mackay, Medir, and Thornton (1975) showed that the oil had a strong negative spreading tendency under ice, that is, oil behaves at an ice-water interface in the same fashion as mercury behaves on a clean glass plate or water on a wax surface. The equilibrium thickness of oil droplets is about 0.7-1.2 cm. As the slick becomes thinner, surface tension forces resist further spreading and an equilibrium is achieved. Since the buoyancy forces are driven by the density difference between the oil and the water which is of the order of 0.15 g/cm^3, compared to 0.85 gm/cm^3 for oil and air, the equilibrium thickness will be considerably larger than for oil spilled on water. This is a favourable effect since it will reduce the area affected by an oil spill under ice.

Experiments by Wolfe and Hoult (1974) are in agreement with this in showing that the equilibrium thickness for oil under ice will be between 0.25 and 1.3 cm. Since the undersurface of the sea ice is quite irregular, the oil will form pools thicker than the equilibrium value and it is likely that the average oil thickness will be about one cm. It is possible that large ice protuberances such as ice keels may dominate the situation in determining the total area of a large spill and will give rise to oil pool thicknesses well in excess of 1 cm. The frequency and configuration of such large topographical features is clearly of considerable importance in estimating the area affected by a spill under ice. A statistical study of this is being conducted as part of the Beaufort Sea program.

The observation that oil does not "wet" ice at an ice-water interface suggests that oil should be relatively easy to remove from under an ice surface with the possible exception of oil which has entered brine channels. This, to some extent, is in conflict with the work of Wolfe and Hoult (1974) who estimated an amount of oil which "adheres" to the undersurface of the ice. They showed that oil does not become substantially trapped into the ice brine matrix on the underside of the sea ice even when ice porosity is high, although some penetration of oil into brine channels is inevitable. As the ice sheet continues to grow, it will physically trap the oil as a sandwich or lens and there will be little or no transfer of water or ice through the oil lens. Wolfe and Hoult have considered the heat transfer problem of ice growth rates and found that it will be of the order of several cm per week and will not be greatly altered by the presence of oil lenses. The oil will thus become inaccessible to simple removal from beneath the ice within a few days after the spill. This has been confirmed experimentally by Bell (1974) at Resolute Bay. Laboratory studies by Keevil and Ramseier (1975) have also demonstrated this behaviour.

Recent studies (Brown 1975) in the Beaufort Sea have shown that oil trapped as a lens between ice is subject to negligible weathering or dissolution. It can be expected to have substantially its original properties and composition when released by thawing in the spring.

It should also be recognized that oil and gas released under ice will cause the ice sheet to lift and even crack, thus leading to oil flow onto the ice surface. If the ice sheet is one or more metres thick, it may be strong enough to support the oil and gas buoyancy but thinner sheets will probably be ruptured.

First-year ice will trap the oil in lenses and little movement up through the ice in winter would be expected. Bell noticed movement of oil up into brine channels from an oil lens. However, in the spring as the ice melts and becomes more porous it is possible that oil may migrate to the ice surface. This would not be expected to occur in multiyear ice, which is mostly fresh ice and contains few brine channels.

The long-term migration of oil through ice has been discussed by Ramseier (1972) who has stated that if oil is left below multiyear pack ice, it will migrate in about four years to the surface through annual freeze-thaw cycles. The rate of normal weathering or aging of the oil may be negligible under these conditions. Evaporation, which normally is the mechanism which accounts for the loss of the more toxic and more volatile compounds in oil will be prevented since the oil is not exposed to air. In addition, freezing under the oil will shut off any dissolution process. Biodegradation will presumably be negligible at the low ice temperatures.

Although no attempt is made here to consider the biological effects in detail, a few observations are appropriate. Meguro et al. (1966) showed that most primary production in polar seas occurs in the 30 cm of water immediately below the ice surface, i.e. at the location where an oil spill would settle. The opacity of the oil will prevent photosynthesis. Sea mammals which depend on leads or breathing holes for air supply will find these locations flooded with oil. There can be little doubt that a combination of toxicity, physical oiling, and lack of light and oxygen will have a devastating effect on marine life under a spill. To what extent surrounding areas will be affected is presently unknown.

Several authors have hypothesized on the extent of the large-scale interactions between oil and ice which may occur (Ramseier 1975, Campbell and Martin 1973, Ayers and Glaeser 1974) and opinions differ as to the extent to which oil could ultimately spread. Estimates vary from a few square kilometres to hundreds of square kilometres for a spill of 2,000,000 barrels of crude oil. It has been also suggested that oil could spread for hundreds of kilometres due to the dynamics of the Beaufort Gyre, as discussed previously.

Oil in Ice-Infested Waters

Perhaps the most difficult situation will be the cleanup of oil in ice-infested waters. In ice-free waters modifications of techniques developed in temperate regions will probably be applicable. In ice-covered waters the ice provides a convenient working platform. In ice-infested waters, the dynamic nature of the ice floes, the large number of mechanisms by which oil can spread under and over ice, the difficulty of deploying craft, and the overriding consideration of personnel safety will make cleanup under these conditions particularly difficult and hazardous. Possibly the most hazardous seasons will be during the spring thaw and fall freezeover. Little or no drilling activities are planned for these periods. However, the edge of the ice pack can move into areas planned for drilling in open water and there is a possibility of oil being released from a prolonged blowout during this period. It should be noted also that the hazardous marine conditions in which floating ice is present may be the cause of a blowout through interference with normal drilling operations. Thus it would appear that consideration of the behaviour of oil under such situations is particularly needed.

There is continuous deformation, expansion, and compression within an ice sheet. This often results in formation of leads of open water between sections of ice and rafting and pressure ridge formation. Oil will be forced from under the ice into leads resulting in a spreading mechanism termed by Campbell and Martin (1973) as "lead-matrix pumping". Since alternate convergence and divergence of ice sections can occur in periods as short as one day, such motions could have considerable effect on oil containment and cleanup operations. While leads do provide areas for oil recovery, their opening and closing and attendant pressure ridge formation will render work on the ice sheet both difficult and hazardous.

Shearing within the ice pack may also result in further dispersion of the oil by the smearing of one section on another, although it may be that the thickness of the under-ice layer of oil would not be decreased by such shear forces and thus the oil would move as a fixed pool with respect to the ice sheet.

The degree of dispersion and the emulsification of oil into a water column below the ice has also been estimated. Campbell and Martin (1973) believe that the experience of the Chedubucto Bay spill (with Bunker C oil) and that at Deception Bay (with diesel oil) described by Ramseier (1973) indicate that significant amounts of oil, perhaps 10% of the total, may be dispersed into the water column. Ayers and Glaeser (1974) felt, however, that there would be insufficient mixing in ice-infested areas where wave and tidal action are small to cause this degree of dispersion and emulsification. They suggested a maximum of 1% of the oil spill would enter the water column below the ice. Although this is a small fraction of the total, it can represent extensive contamination of the water column. A layer of oil one metre square by 1 cm thick will have a mass of about 8500 g. If 1% or 85 g of this is dispersed in water at a level of 1 mg/1 (20 times the accepted minimum level of 0.05 mg/1 at which oil has sub-lethal effects) then there is sufficient oil to contaminate a column of water 85 m deep. It is apparent that in turbulent ice-water-oil situations sufficient dispersion may occur to cause a severe level of contamination to an appreciable depth. The transport rates in such a situation have been considered by Leinonen and Mackay (1975) but unfortunately there is little information on the values of vertical eddy diffusivities in such regimes and predicted rates are very sensitive to the values chosen. Mackay and Shiu (1975) have also discussed this issue.

Cleanup

The Nature of the Spill

The design and testing of a system to allow containment, recovery, and disposal of oil from a blowout presents a difficult technical problem. It is assumed here that the situation to be faced is an oil flow rate of up to 800 m^3 (5000 barrels) per day. This could be in conditions ranging from calm open seas to solid pack ice to floating ice with open leads. Assuming that it requires five days to implement successful containment, approximately 4000 m^3 of crude oil would be in place on the water or on or under the ice. This would cover an area of perhaps 0.4 km^2 based on a film thickness of 1.0 cm. In practice it seems likely that an area possibly from one-tenth to one-half of this is likely to be affected due to interference of oil flow by ice keels.

Assuming that the oil from a blowout will flow at roughly the same rate for up to a period of one month, it is apparent that the total quantity of oil to be handled is of the order of 24,000 m^3 contained in an area 0.5 to 2 km^2.

To date, no new techniques or equipment have been devised and tested to allow such a situation to be handled. The approach taken by government and industry has generally been to use cleanup methods which have been demonstrated to be effective in southern Canada and modify them for Arctic conditions. There is understandably some reluctance on the part of government to allow large quantities of oil to be spilt deliberately in order to test oil spill clean-up equipment. However, experience gained from such oil spills would be immensely valuable and the resultant environmental damage will be small in comparison to the mitigation of the environmental damage of a large spill.

Summer open-water situations are likely to be handled using conventional equipment proven for clean-up in temperate waters. Winter situations are likely to be handled in a manner analogous to that of a winter terrestrial spill in southern Canada. There is at present an almost complete lack of preparedness to handle spills of oil under ice or in ice-infested waters. The equipment which is presently available in the Mackenzie Delta Environmental Protection Unit (a co-operative of six oil exploration companies) is similar to that maintained in Alberta and Ontario.

The approach which is suggested here is to use whenever possible proven temperate techniques for clean-up, with modification for cold conditions where necessary or feasible, and to develop new techniques for the ice-covered and ice-infested conditions. It seems likely that oil on water will be contained and recovered by traditional methods. These are reviewed briefly here. The more challenging problem of recovering oil from under and between ice is also considered. Regardless of the measures adopted, it is essential that a contingency plan be prepared well in advance. This should contain the usual components of notifying the relevant agencies, obtaining extra personnel (who have been suitably trained), arranging accommodation and transportation, moving equipment to the site, deploying containment systems, oil removal devices, and oil storage and disposal facilities. A full knowledge of the current and wind characteristics in the area and the way in which the oil will probably behave, especially the direction of drift, are essential. It will also be necessary to have available equipment for rapid ice penetration in order to locate oil under the ice. Locating the oil is usually easy on water or land but under ice or snow a systematic search may be required to determine the extent of oil movement.

Conventional Clean-up Measures

Glaeser and Vance (1971) tested various temperate oil spill control measures under Alaskan summer conditions and McMinn and Golden (1972) tested similar equipment under Alaskan winter conditions. Glaeser and Vance showed that peat and straw were quite useful for absorbing oil spilled on cold water at 5°C and on ice and data were reported on the absorptive capacity of both materials. Difficulties were experienced with peat moss and with commercial absorbents, both in spreading because they were easily blown by wind and in collecting due to the fluid nature of the oil-saturated mass.

The winter study results were less encouraging since both natural and artificial absorbents were ineffective because of difficulties in applying them under windy conditions and because considerable mixing was required to bring the oil in intimate contact with the absorbent. Presumably the higher viscosities associated with lower temperatures result in a reduced rate of penetration of the oil into the absorbent. Absorbents are only useful for removing small quantities of oil, i.e. approximately their own volume or 10 to 50 times their mass. The magnitude of the spill is such that absorbents can only be used for clean-up of a small fraction of the spill, the bulk of the oil being removed by other means.

Dispersants were ineffective when added to oil on ice since they served only to decrease the oil viscosity slightly. McLean (1972) stated that dispersants were also ineffective when used on oil slicks in near freezing temperatures. It is probably best to avoid dispersion rather than encourage it since the biological problem is worsened by widespread oil dispersion. The exception to this may be a situation where the major environmental concern is gross physical oiling of mammals or birds and break-up of the oil slicks may reduce the amount of such oiling.

Surface active agents (herders) were predicted by McMinn and Golden to fail under very cold conditions because oil spreading on ice is not influenced significantly by surface tension flows. This was confirmed by experiment. Such agents would be largely ineffective and unnecessary for oil spreading under ice because surface tension opposes spreading under these conditions and the herder would have little or no additional effect.

The use of either commercial or makeshift mechanical containment booms for placement through the ice sheet has not been attempted in the Arctic, although some small fresh water applications in real spill situations have taken place in Ontario, Alberta, and in the U.S. In these situations, the oil was contained but it is uncertain whether any flow of oil under the ice would have occurred had the booms not been placed. McLeod and McLeod (1972) state that several manufacturers have successfully tested booms in open water at sub-freezing temperatures. Field trials were conducted by Getman (1975) in Cook Inlet, Alaska, to determine the effectiveness of various commercial oil booms in ice-infested water. The study did not include the use of oil. The results indicated two modes of boom failure caused by moving ice. The first, which was for external tension member booms, was the riding of the boom on top of the ice as the ice floe passed. The second, for integral tension member, heavy duty booms, was breakage of the boom at a connector point, or tearing of the fabric. It is possible that a boom which fails temporarily by not resisting the ice floe may have some application.

It appears, however, that the use of booms under conditions of moving ice as present in the Beaufort Sea pack ice zone is impractical unless channels can be kept free of ice. There is obviously a need for development of robust booms which retain their flexibility in cold temperatures. Consideration should be given to permanent deployment of booms and a submerged tent-like arrangement around drill ships. This could greatly mitigate the effects of an oil release under ice-free conditions. Such permanent booms are presently used in marine terminals.

The only temperate method other than burning which has proved successful in near-freezing temperature waters and in ice-infested waters has been surface skimming of oil. McLean (1972) reported that at Chedabucto Bay a "slick-licker" or rotating absorbent belt skimmer was used successfully in near-freezing sea water. Ramseier et al. (1973) in their discussion of the Arctic diesel oil spill at Deception Bay, Quebec, reported that a floating surface skimmer was used successfully to pump oil from pools on top of and between the ice. Getman (1975) conducted trials of various skimming devices under relatively mild (15°F) ice-infested water conditions. Again, no oil was used. However, an absorbent belt skimmer and an adjustable weir skimmer both appeared to have potential applicability.

At low temperatures, these devices may be subject to icing and very careful design is necessary to ensure trouble-free operation. The required capacity will be such that only the large throughput versions will be capable of handling a significant fraction of the spilled oil.

Burning as a means of removal or disposal of spilled crude oil offshore in the Beaufort Sea appears to be feasible. Field experiments conducted by Glaeser and Vance (1971), McMinn and Golden (1972) in the Arctic, and Greene and Mackay (1975) in Ontario, indicate that fresh crude oil at sub-freezing air temperatures can be successfully burned. Crude oil which has been under ice for several weeks and crude oil which has been weathered on top of ice for a number of days can both be successfully ignited with a small hand-held torch. Burning agents or combustion promoters have not proved to be necessary to improve the rate or degree of burn achieved. Successful burns have been conducted for oil slicks both on ice and on cold water in trenches through the ice.

The studies also have indicated that oil slick thickness must be at least 0.5 cm for successful combustion. Thinner slicks ignite but burn for a relatively short period of time, presumably because of the more effective cooling by the underlying ice. Burn efficiency experiments indicate that 80-90% of the crude oil pool can be burned. The residue is a waxy black skin of about 1 mm thickness floating on top of water in the trench or on the pool of water which forms on the ice during a burn. This residue is easily skimmed off and could presumably be stored and ultimately disposed of. It is essential that such an operation be conducted before fresh or blowing snow covers the slick.

A potential problem with burning is the formation of dense black smoke which is associated with very violent burns. There may be a fallout of black soot particles over a wide area and this may reduce the albedo of the ice causing a greater absorption of radiation and hence an early melt.

It should be noted that a burn of crude oil is very violent for about three-quarters of the burn time and this requires that extreme caution be practised when burning. The ignition of the slick is particularly dangerous since there may be an explosion hazard with an attendant danger to personnel.

The use of biodegradation methods to destroy oil has been considered but approaches such as innoculation with specially developed "oil eating" bacteria are now viewed with scepticism. Arctic temperatures are low causing slow degradation and nutrients may be in limiting supply. Introducing new strains of bacteria to an environment may cause new problems and it is probably best to allow the indigenous strains to adapt and consume oil at whatever rate local conditions dictate, recognizing that the oil may take many years for ultimate degradation.

Sinking agents have also fallen into disrepute in recent years since they merely serve to solve the aesthetic problem and aggravate the biological problem severely by exposing benthic organisms to oil.

It is interesting that the strategy adopted by the U.K. North Sea Operators is to store large quantities of low toxicity dispersants for treating offshore spills. An open seaboom system (BP Sea-Pack) is being added to the equipment available to permit containment and ultimately recovery of the oil (Lester and Beynon 1973).

Cleanup from Under Ice

This problem can be broken down into three aspects, penetration of the ice sheet to reach the oil, inducing the oil to flow from under the ice to some storage and possibly oil-water separation facilities, and final oil disposal. It is first prudent to examine the advantages which accrue to the clean-up operation by having oil in this environment and to devise methods of using these advantages to the maximum.

First the ice (if thick enough) provides a convenient working platform for clean-up operations and thus the use of heavy equipment can be considered. Second the recovered oil has a high calorific value, thus there is a ready source of heat which is always advantageous in a cold climate. Third, the oil will be less mobile since it will not be subject to wind movement, thus it may be easier to contain.

Ice penetration can be accomplished by mechanical means such as augers, saws, or trenching devices. The large pipeline trenchers which have been tested by Panarctic for pipelaying from ice are probably too expensive and immobile for use in clean-up measures. The most attractive device is judged to be a gasoline-powered sledge-mounted auger, towed by snowmobile. It would be sensible to concentrate drilling on the down-current side of the spill.

Blasting is another approach to penetration, however it does introduce some biological, fire, and personnel hazards, requires skilled handling, and will require bore holes. It must be regarded as a viable alternative to mechanical penetration.

Melting may also be used, especially if means can be devised to obtain the heat required by burning the recovered oil. A supplementary fuel source would be necessary prior to oil recovery. If hot water could be generated in a furnace this could be used to melt a trench or if pumped down bore holes it would melt the ice at that ice-water interface forming a "high" ice level which could be used to contain and recover the oil. The currents generated by the water flow would also assist the oil flow to the surface.

Another application of melting would be to keep trenches open by pumping hot water into them or by floating a plastic pipe containing circulating hot water in the trench.

Oil can be contained either by inserting a boom which requires cutting a slot or trench or by creating a high point in the ice-water interface such as a trench or lead or melted area. If keels are naturally present these could be used for containment. The only method of locating them is probably by the use of divers. Indeed, divers may form an invaluable part of any clean-up team.

One longer-term method of containment may be to build up a thickened ice sheet by pumping water onto the ice surface and allowing it to depress the ice-water interface. Such techniques have been used for making thick ice island drilling platforms by Panarctic. Greene and Mackay (1975) in a field-scale experimental spill of crude oil under fresh water ice showed that when a hole or trench was cut through the ice into an oil layer at the ice-water interface, only the oil under the trench and within a few cm perimeter flowed up the hole. In a blowout situation, the continuing volume of oil rising on the ice sheet will cause flow along the bottom of the ice under the influence of buoyancy forces. However, should the oil spread to a thickness of 1 cm (for a smooth under-ice topography), there will be no net force to move oil into trenches for recovery. In such a situation oil displacement into trenches would require some means of inducing oil flow along the ice subsurface. Greene and Mackay (1975) were able to cause sufficient agitation of the oil at the ice-water interface by pumping compressed air under the ice sheet, to force the oil to flow into the trenches from where it could be recovered by means of pumps and skimmers. A gasoline powered compressor could be used for this purpose. Other means of producing a current would be pumping hot water or steam below the ice, and in fact these could be produced by incineration of recovered oil. Open leads in the ice sheet could serve well, in the place of artificial trenches, for oil recovery.

An alternative to recovery of oil as it flows into trenches or leads by agitation under the ice sheet, is in situ burning of the oil as it surfaces and collects in trenches. If compressed air is used as the means for inducing flow, this air would surface with the oil and aid combustion.

Assuming that the oil is not burned in situ the best method of oil removal is by pumping. Inevitably some water will be pumped with the oil and this should be separated later, most likely by a gravity settling system, the oil going to storage or disposal and the water back to the sea. Such a system would have to be carefully designed to prevent freezing.

It is feasible to provide oil storage by bladder tanks for several thousand gallons of oil but since the total quantity released may be millions of gallons, tank storage is at best a temporary measure. Transportation by pipeline or barge to shore is not likely to be feasible. Thus the remaining alternatives are storage in a diked area on the ice or disposal by burning.

If the ice sheet is thick and stable a storage area could be created by forming ice or snow dikes to contain the oil until removed by other methods. Dikes could be constructed by bulldozers building snow banks or by pumping water to form ice barriers. Large plastic liners could be useful for retaining oil in diked areas. It should be noted that the fire hazard from such areas will be considerable.

The best oil disposal method is burning using specially designed incinerators which will process oil at flow rates roughly equivalent to the oil spill rate without creating a smoke-soot problem. However, if oil from beneath the ice comes to the surface during spring thaw, either by being buoyed up through brine channels in first-year ice, or being pumped through leads or cracks in multiyear ice and first-year ice, it may be possible to burn the oil in situ as it reaches the surface. This would by-pass the necessity of using skimmers, pumps, and containment vessels which are attendant on incineration.

Conclusions

The present status of plans to drill offshore in the Beaufort Sea has been discussed and it is suggested that the presence of ice, adverse weather, and high formation pressures will contribute to more hazardous drilling conditions than are generally experienced offshore. A brief analysis of blowout incidents suggests about a 0.5 to 1% frequency in this area with volumes of 60,000 to 150,000 barrels of oil released in the first month and several times these volumes in total. The area affected would be of the order of 0.5 to 2 km^2. The physical behaviour of oil rising and spreading in Arctic waters, spreading on and under sea ice and in ice-infested waters has been reviewed. Clean-up approaches have been discussed and it appears that new techniques are necessary for oil retrieval, but that burning is the best method of disposal. The containment, recovery, and disposal of spilled oil under Beaufort Sea conditions presents a severe technical problem. Little is currently known about the physical and biological behaviour of such spills. To proceed to drill without adequate background knowledge of the environment or an adequate clean-up capability is to accept the risk of a substantial uncontrollable oil spill incident which could have severe environmental and social impacts.

References

Alpine Geophysical Associates Inc., *Oil pollution incident Platform Charlie; Louisiana*, EPA Project #15080 FTU, May 1971.

Ayers, R.C., Jahns, H.O., and Glaeser, J.L., *Oil spills in the Arctic Ocean: extent of spreading and possibility of large-scale thermal effects, Science*, vol. 186, November 1974.

Bell, L., Water Associates Ltd., Tobermory, Ontario, personal communication from the Arctic IV expedition to Resolute Bay, N.W.T., 1974.

Brockis, G.J., *Industry emergency oil spill plans and programmes,* paper presented to Institute of Petroleum meeting, November 1974.

Brown, R.F., *The interaction of crude oil with Arctic sea ice,* paper presented at the third International Conference on Port and Ocean Engineering, University of Alaska, August 1975.

Brown, R.F., *Investigation and analysis of September 1970 storm — Beaufort Sea,* Public Works of Canada Report, January 1971.

Campbell, W.J., and Martin, S., *Oil and ice in the Arctic Ocean: posssible large-scale interactions, Science,* vol. 18, July 1973.

Canada, Department of Supply and Services, *Methods and techniques of oil disposal on a large scale under Arctic conditions,* Contract No. SS01-KE204-4-EP12, 1974.

Chen, E.C., Overall, J.C.K., and Phillips, C.R., *Spreading of crude oil on an ice surface, Can. J. Chem. Eng.,* vol. 52, February 1974.

Dunbar, M., and Wittman, W., *Some features of ice movement in the Arctic Basin,* in *Proceedings of the Arctic Basin Symposium* (Montreal: Arctic Institute of North America, 1962).

Glaeser, J.L., and Vance, G.P., *A study of the behavior of oil spills in the Arctic,* U.S.C.G. Office of Research and Development Report on Project #714108/A/001, 002, February 1971.

Getman, J., *United States Coast Guard Arctic oil pollution control program,* in *Proceedings of Oil Pollution Prevention and Control Conference,* San Francisco, 1975.

Greene, G.D., and Mackay, D., unpublished report on the behavior of an oil spill under fresh water ice at Baie du Doré, Ontario, Institute for Environmental Studies, University of Toronto, 1975.

Kash, D., et al. *Energy Under the Oceans: A Technology Assessment of Outer Continental Shelf Oil and Gas Operations* (Norman, Oklahoma: University of Oklahoma Press, 1973).

Keevil, B.E., and Ramseier, R.O., *Behavior of oil spilled under floating ice,* in *Proceedings of Conference on Prevention and Control of Oil Pollution,* San Francisco, 1975.

Leinonen, P.J., and Mackay, D., *Mathematical model of evaporation and dissolution from OIIL SPILLS ON ICE, LAND, WATER AND UNDER ICE,* IN *Proceedings of the 10th Canadian Symposium of Water Pollution Research,* Toronto, February 1975.

Lester, T.E., and Beynon, L.R., *Pollution and the offshore oil industry, Marine Pollution Bulletin,* February 1973.

Logan, W.G., *Oil spill countermeasures for the Beaufort Sea,* in *Proceedings of Conference on Prevention and Control of Oil Pollution,* San Francisco, 1975.

Mackay, D., and Shiu, W.Y., *The aequeous sollubility and air-water exchange characteristics of hydrocarbons under environmental conditions,* in *Chemistry and Physics of Aqueous Gas Sollutions,* Electrochemical Society, 1975.

Mackay, D., Medir, M., and Thornton, D., *The interfacial behavior of oil under ice,* unpublished report of the Institute for Environmental Studies, University of Toronto, 1975.

Mackay, D., et al. *The behavior of crude oil spilled on snow, Arctic,* vol. 28, March 1975.

Markham, W.F., Ice Climatologist, Atmospheric Environment Service, DOE, Ottawa, personal communication, February 1975.

McLean, A.Y., *The behavior of oil spilled in a cold water environment,* in *Proceedings of 4th Annual I.E.E.E. Offshore Technology Conference,* 1972.

McLeod, W.R., and McLeod, D.L., *Measures to combat offshore Arctic oil spills,* in *Proceedings of 4th Annual I.E.E.E. Offshore Technology Confrence, 1972.*

Meguro, H., Kuniyuki, I., and Fukushima, H., *Ice flora (bottom type) — a mechanism of primary production in polar seas and the growth of diatoms in sea ice, Science,* vol. 152, 1966.

Moore, S.F., Duyer, R.L., *Effects of oil on marine organisms, a critical assessment of published data, Water Research,* vol. 8, 1974.

Oil on the Sea, edited by David P. Hoult, in the Ocean Technology Series, edited by John P. Draven (New York: Plenum Press, 1969).

Oilweek, a weekly publication of MacLean-Hunter Ltd., Calgary, Alberta; issue dates are given in the text.

Pimlott, D., *The hazardous search for oil and gas in Arctic waters, Nature Canada,* vol. 3, October-December 1974.

Ramseier, R.O., *Possible fate of oil in the Arctic Basin,* paper presented at First World Congress on Water Resources, Chicago, 1972.

Ramseier, R.O., et al, *Oil spill at Deception Bay, Hudson Strait,* Environment Canada Inland Waters Directorate, *Scientific Series,* vol. 29, 1973.

Ramseier, R.O., Head, Floating Ice Section, Department of the Environment, Ottawa, personal communication, February 1975.

Rogers, J.C., *Arctic Sea ice dynamics,* in *Proceedings of 6th Annual I.E.E.E. Offshore Technology Conference,* vol. 2, 1974.

Rosandhaug, K., *Development of a supply vessel for Arctic oil operations,* in *Proceedings of 6th Annual I.E.E.E. Offshore Technology Conference,* vol. 1, 1974.

Ross, S.L., *Oil spill technology development in Canada,* in *Proceedings of Conference on Prevention and Control of Oil Pollution,* San Francisco, 1975.

University of California, Santa Barbara, *Santa Barbara oil pollution,* 1969, F.W.Q.A. Department of Interior, Program No. 15080, D7R 10/70.

Wickland, R.I., *Significance of polluted pools containing oil and pesticides discovered in the Canadian Arctic,* report by J.A. MacInnis Foundation, May 1974.

Wolfe, L.S., and Hoult, J.P., *Effects of oil under sea ice, J. Glaciology,* vol. 13, 1974.